Maths A–Z

Editor: P J F Horril

Longman

Longman Group Limited,
Longman House, Burnt Mill, Harlow,
Essex CM20 2JE, England
and Associated Companies throughout the world.

© Longman Group Limited 1986

First published 1986

British Library Cataloguing in Publication Data

Horril, P.J.F.
  Dictionary of mathematics.
  1. Mathematics—Dictionaries
  I. Title
  510′.3′21   QA5

0-582-55588-4

Set in Linotron 202 Times New Roman
by Computerset (MFK) Ltd., Ely, Cambridgeshire.
Printed in Great Britain
by Richard Clay PLC, Bungay.

Contents

Foreword

This guide to the many specialized terms used in modern mathematics is intended chiefly for those studying the subject at school or college, but should prove useful to students of all science subjects and everyone whose work brings them into contact with this field.

The dictionary contains explanations of over 1000 terms and concepts encountered in public examinations in mathematics. The definitions are fully supplemented by formulae, worked examples, and over 100 diagrams, making this book invaluable for handy reference and revision.

Cross references to other entries in the dictionary are shown in small capitals.

Abbreviations used in this dictionary

abbr	abbreviation
b	breadth
C	Celsius, Centigrade
cm	centimetre
d	diameter
esp	especially
F	Fahrenheit
ft	foot, feet
g	gram
h	height
in	inch
kg	kilogram
km	kilometre
l	length
lb	pound
m	metre
mm	millimetre
r	radius
s	second
specif	specifically
usu	usually
yd	yard

A list of mathematical symbols is given at the back of the book.

abacus (*plural* **abaci** *or* **abacuses**)
an instrument for performing calculations by sliding bead counters along rods or in grooves

Abel, Niels (1802–29) Norwegian mathematician noted for his work in modern algebra, esp group theory

Abelian
COMMUTATIVE 2 ⟨~ *group*⟩ ⟨~ *ring*⟩

abscissa (*plural* **abscissas** *also* **abscissae**)
the coordinate of a point in a Cartesian coordinate system obtained by measuring parallel to the *x*-axis; the *x*-coordinate of a point in a plane – compare ORDINATE

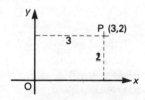

The abscissa of point P is 3

absolute value
MODULUS 1, 2

acceleration
increase in speed or velocity; *specif* the rate of change of velocity with respect to time

> **Example** If an object increases its speed from 10 metres per second (ms^{-1}) to $50ms^{-1}$ in 10 seconds while moving in a straight line, then its (average) acceleration is 4 metres per second per second (written $4ms^{-2}$).
>
> If a vector v represents the velocity of a point P, then its (vector) acceleration is dv/dt.

accuracy
freedom from error; correctness. The degree of accuracy is an indication of the error involved in an approximation
(eg length = 3.4 cm, correct to 2 significant figures, means that $3.35cm \leq$ length $< 3.45cm$).

acre
a unit equal to $4840yd^2$ ($4046.86m^2$)

actuary
a statistician who calculates insurance risks, premiums, annuities, etc

acute
1 *of an angle* measuring between 0° and 90°
2 composed of acute angles ⟨*an ~ angled triangle*⟩

addition
an act, process, or instance of adding; *esp* the operation of combining numbers so as to obtain a single equivalent number

addition formulae
formulae used in trigonometry to find the sine, cosine, or tangent of the sum or difference of two angles:

$\sin (A + B) = \sin A \cos B + \cos A \sin B$
$\sin (A - B) = \sin A \cos B - \cos A \sin B$
$\cos (A + B) = \cos A \cos B - \sin A \sin B$
$\cos (A - B) = \cos A \cos B + \sin A \sin B$

$$\tan (A + B) = \frac{\tan A + \tan B}{1 - \tan A \tan B}$$
$$\tan (A - B) = \frac{\tan A - \tan B}{1 + \tan A \tan B}$$

– compare DOUBLE ANGLE FORMULAE, FACTOR FORMULAE

additive
of or characterized by addition

adjacent
1 being one of the sides that form a given angle in a triangle. In a right-angled triangle it is the side that with the hypotenuse makes a given angle.
2 *of two sides of a polygon* sharing a common vertex
3 *of two angles* standing on a straight line; having a vertex and one side in common

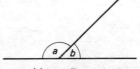

a and *b* are adjacent angles

adjoint
the transpose of a matrix in which each element has been replaced by its cofactor

affine
of or being a transformation that preserves straightness and parallelism of lines but may alter distances between points and angles between lines ⟨*~ geometry*⟩

aleph-null *also* **aleph nought**
the cardinal number of the set of natural numbers {1, 2, 3, 4, etc}, usu indicated by the symbol χ_0

algebra

1 a branch of mathematics in which letters, symbols, etc representing numbers, sets of numbers, or similar entities are manipulated according to special rules of operation
2 a system of representing logical arguments in symbols

algebraic

1 relating to, involving, or according to the laws of algebra
2 involving only a finite rather than an infinite number of repetitions of addition, subtraction, multiplication, division, extraction of roots, and raising to powers ⟨*an ~ equation*⟩ – compare TRANSCENDENTAL

algebraic number

a number that can be the solution of a polynomial equation with rational coefficients – compare TRANCENDENTAL NUMBER

algorithm

a systematic procedure for solving a mathematical problem in a limited number of steps, usu involving the repetition of a single operation several times; *broadly* a step-by-step procedure for solving a problem or accomplishing some end

> **Example** An algorithm for finding $\sqrt{17}$.
> Let a be an approximate value of $\sqrt{17}$, then
> $$\tfrac{1}{2}(a + 17/a)$$
> is a better approximation.
> If we take $\sqrt{17} \approx 4$ as a first approximation, then the second is
> $$\tfrac{1}{2}(4 + 17/4) = 4 \cdot 125.$$
> Repeating the algorithm gives $4 \cdot 123$; the process may be repeated until the required degree of accuracy is obtained.

aliquot

of a divisor or part contained an exact number of times in a larger whole (eg 5 is an aliquot part of 15) – see also FACTOR

alternate angles

a pair of angles on opposite sides of a transversal which intersects two other lines. In the diagram, *a* and *b* are alternate angles. If AB and CD are parallel, the alternate angles are equal.

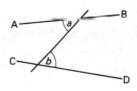

altitude

the perpendicular distance from one side of a geometric figure (eg a triangle or cone) to the opposite vertex; *also* a line drawn on a geometric figure

corresponding to such an
altitude

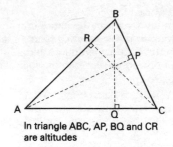

In triangle ABC, AP, BQ and CR
are altitudes

amplitude
the maximum displacement of a periodic function from the
horizontal axis

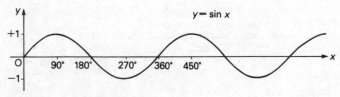

The amplitude of the function sin x is 1

analysis
a branch of mathematics concerned with the rigorous treatment of
the ideas of limits, functions, calculus, etc

analysis of variance
analysis of variation in an experimental outcome, esp a statistical
variation (variance) in order to determine the contributions of
given factors or variables to it

analytic *or* analytical
1 treated or solvable by or using the methods of mathematical
analysis ⟨*the ~ solution*⟩
2 *of a function of a complex variable* being defined and having a
derivative at a specified point or at all points of some region

analytical geometry
the study of geometric properties by means of algebraic operations
on coordinates in a coordinate system – called also COORDINATE
GEOMETRY

and
a logical connective denoted by ∧

angle
the difference in the directions
of two intersecting lines; the
amount of rotation (normally

measured in degrees or
radians) required to rotate one
of a pair of intersecting lines,
about the common point, onto
the other line

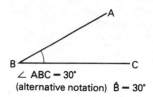

∠ ABC — 30°
(alternative notation) B̂ — 30°

angle of depression
the angle formed by the line of sight and the horizontal plane for
an object below the horizontal. The angle of depression α, for an
observer O who is above the level of an object P, is the angle
between the horizontal line through O and the line OP.

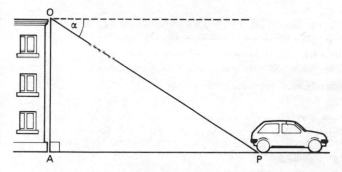

angle of elevation
the angle formed by the line of sight and the horizontal plane for
an object above the horizontal. The angle of elevation α, for an
observer O who is below the level of an object P, is the angle
between the horizontal line through O and the line OP.

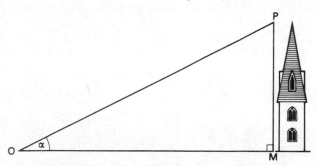

annulus
the region bounded by a pair of
concentric circles.

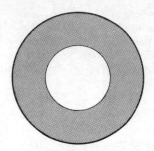

antiderivative
INDEFINITE INTEGRAL

antilog
an antilogarithm

antilogarithm
the number whose logarithm is a given number (eg the logarithm
(base 10) of 1000 is 3, the antilogarithm (base 10) of 3 is 1000)

Apollonius, (of Perga) Greek mathematician, approximately 3rd
century BC, who wrote a treatise on conic sections

approximate
nearly correct or exact (eg 22/7 is an approximate value of π)

approximation
a mathematical quantity that is close in value but not equal to a
desired quantity (eg 1·73 is an approximation to $\sqrt{3}$)

arabic numeral
any of the number symbols 0, 1, 2, 3, 4, 5, 6, 7, 8, 9

arc (*noun*)
a portion of a curve (eg part of
a circumference)

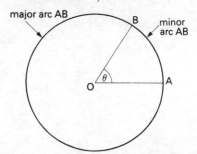

If $\angle$AOB $= \theta°$, the length of the minor
arc AB is $\dfrac{\theta}{360} \times 2\pi r$

arc (*adjective*)
mathematically inverse – used of inverse trigonometric functions and hyperbolic functions (eg arc sin $0 \cdot 5 = 30°$; alternative notation: $\sin^{-1} 0 \cdot 5 = 30°$)

Archimedes, Greek mathematician, inventor and astronomer, approximately 3rd century BC

are
a metric unit of square measure equal to 100m²

area
1 the surface included within a boundary, geometrical drawing, set of lines, etc. The area of some simple figures can be calculated from a formula:

$$\text{square } l^2$$
$$\text{rectangle } bh$$
$$\text{triangle } \tfrac{1}{2}bh$$
$$\text{circle } \pi r^2$$
$$\text{sphere } 4\pi r^2$$

2 the extent of a surface as measured in an equivalent number of squares of standard size ⟨*the ~ of the tabletop is 1·5m³*⟩

Argand, Jean (1768–1822) Swiss mathematician. Originator of the Argand diagram.

Argand diagram
a graph consisting of two axes, on which a complex number $x + iy$ (where $i = \sqrt{-1}$) can be represented, the x-axis being taken to be the real axis, and the y-axis being taken to be the imaginary axis

argument
the argument of a complex number $x + iy$ is the angle in the Argand diagram between the real axis and the line joining the origin to point (x,y), ie arc tan (y/x)

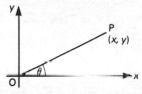

Angle θ is the argument of $x + iy$

arithmetic
1 a branch of mathematics that deals with the operations of addition, multiplication, subtraction, and division
2 skill in or calculation using numbers

arithmetic mean
the value found by dividing the sum of a set of terms by the number of terms. The arithmetic mean of the n numbers $x_1, x_2, x_3, ..., x_n$ is $(x_1 + x_2 + x_3 + ... + x_n)/n$ (eg the arithmetic mean of 4, 5, and 9 is 6).

arithmetic progression (*abbr* **AP**)
a sequence (eg 3, 5, 7, 9) in which each number differs from the preceding one by a constant amount. The sum of an AP with n terms whose first term is a and last term is l is $\frac{1}{2}n(a + l)$.

array
an arrangement of mathematical or statistical data in a specific order; *specif* an arrangement of numbers, symbols, or other mathematical elements in rows and columns – see MATRIX

associative
of or being a binary mathematical operation (eg addition or multiplication) such that $(x * y) * z = x * (y * z)$, where $*$ denotes the operation (eg $(2 \times 3) \times 5 = 2 \times (3 \times 5)$)

assumption
a fact or statement (eg a proposition, postulate, or established principle) taken to be true without proof

asymptote
a straight line which is approached more and more closely by a curve such that the tangent to the curve comes nearer and nearer to coinciding with the line, but never intersects it

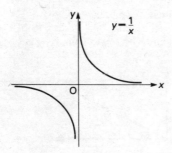

The lines $x = 0$ and $y = 0$ are asymptotes

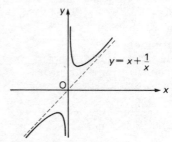

The lines $x = 0$ and $x = y$ are asymptotes

average (*noun*)
a single value (eg a mean, mode, or median) that summarizes or represents the general significance of a set of unequal values; *esp* ARITHMETIC MEAN

average (*adjective*)
equalling an arithmetic mean

avoirdupois weight *or* **avoirdupois**
the series of units of weight based on the pound of 16 ounces and
the ounce of 16 drams

axiom
a postulate (eg "the shortest distance between two points is a
straight line")

axis (*plural* **axes**)
1 a straight line with respect to which a body or figure is
symmetrical

 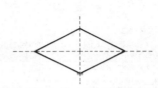

The axis of symmetry of a cone   The (two) axes of symmetry of a rhombus

2 a straight line about which a line, curve, or plane figure revolves
in generating a solid of revolution
3 any of the reference lines of a system (eg a graph) for denoting
position on a plane or in space

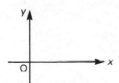 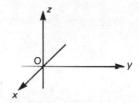

The x- and y- axes in a (two
dimensional) plane. The
equation of the x-axis is $y=0$;
the equation of the y-axis is
$x=0$

The x-, y- and z- axes in a
(three dimensional) space. The
x-axis comes *out* of the page
and is perpendicular to each of
the other axes

azimuth
horizontal direction expressed as the angular distance between the
direction of a fixed point (eg an observer's heading) and the
direction of an object; the angular distance between the vertical
circle of an object (eg a celestial body) and a fixed vertical circle

B

Babbage, Charles (1792–1871) Cambridge mathematician and pioneer of computing who invented the "analytical (or difference) engine", a mechanical device for performing arithmetical operations

bar chart
a diagram for illustrating statistical information in which the frequency with which an item occurs is represented by the length of a vertical bar

Barrow, Isaac (1630–77) Lucasian Professor of Mathematics at Cambridge who resigned his Chair in favour of his pupil, Isaac Newton

base
1 the side or face of a geometric figure on which it is considered to stand
2 the number on which a system of counting is constructed. For example, the decimal system uses the base 10:
$$4023_{10} = 4 \times 10^3 + 0 \times 10^2 + 2 \times 10^1 + 3$$
the binary system uses base 2:
$$1101_2 = 1 \times 2^3 + 1 \times 2^2 + 0 \times 2^1 + 1 \ (=13)$$
3 a number with reference to which logarithms are calculated. For example, $10^3 = 1000$, so the logarithm base 10 of 1000 is 3 (written $\log_{10} 1000 = 3$)

base unit
BASIC UNIT

base vector
in an n-dimensional vector space, a set of n independent base vectors forms a *basis*; any other vector can be expressed as a linear combination of the base vectors. For example in a three-dimensional space the (usual) base vectors are $\mathbf{i}, \mathbf{j}, \mathbf{k}$; any other vector can be expressed in the form $x\mathbf{i} + y\mathbf{j} + z\mathbf{k}$.

basic unit
any of the units of measurement (eg the metre, kilogram, or second) on which all the other units of a particular system of measurement are based – compare DERIVED UNIT

basis
– see BASE VECTOR

Bayesian

Bayesian
being or relating to a statistical theory or method, often used in decision-making, in which new information can be incorporated into what is already known about the likelihood of an event occurring, thus allowing the refinement or alteration of a previous estimate or decision

bearing
the compass direction of one point as measured from another (eg one's present position). The term is used in navigation and surveying for describing the direction of an object. The observer starts by facing due North and turns clockwise until he is facing the object; the angle turned through is the bearing. It is normally written as a 3 figure number.

Examples Due East is bearing 090°; Southeast is 135°; Northwest is 225°; due West is 270°

Bernoulli
the Bernoulli family produced many famous mathematicians during the 17th and 18th centuries. Several of them made substantial contributions to the theory of probability.

Bernoulli trial
a statistical experiment that has two mutually exclusive outcomes, each of which has a constant probability of occurrence (eg tossing a coin for heads or tails is a Bernoulli trial)

bias
a tendency of a statistical estimate to deviate from its expected value (eg because of nonrandom sampling)

biased *or* biassed
1 exhibiting or characterized by bias
2 tending to yield one outcome more frequently than others in a statistical experiment $\langle a \sim coin \rangle$

bilinear
of the first degree (not squared, cubed, etc) with respect to each of two mathematical variables (eg $x + y = 4$ is a bilinear equation)

billion
1 (in American usage) a thousand millions (10^9)
2 (in British usage) a million millions (10^{12})

bimodal
having two statistical modes

binary
1 of, being, or belonging to a system of numbers having 2 as its base. In the binary system, an integer is expressed as the sum of

powers of 2, eg

$$13 = 8 + 4 + 1 = 1 \times 2^3 + 1 \times 2^2 + 0 \times 2^1 + 1 \times 2^0$$

which is abbreviated to

$$13_{10} = 1101_2.$$

Only the digits 0 and 1 are used in a number expressed in binary form.

2 relating or combining two mathematical elements (eg multiplication is a binary operation for combining two numbers)

binomial
a mathematical expression consisting of two terms connected by a plus sign or a minus sign

binomial distribution
a probability distribution each of whose values corresponds to the probability that a specific combination of two types of possible event will occur in a given proportion of statistical trials

binomial theorem
a mathematical theorem in which a binomial $x + y$ raised to the nth power (eg if $n = 2$, the binomial is $(x + y)^2$) is written out in a series of $n + 1$ terms, the general term having the form

$$\left[\frac{n!}{k!(n - k)!} \right] x^k y^{(n-k)}$$

biomathematics
the mathematics of biology

biometry *or* biometrics
the statistical analysis of observations made and experiments conducted in biology

bisect
to divide (eg an angle or line segment) into two equal parts

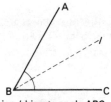

Line *l* bisects angle ABC

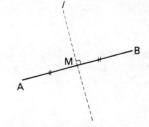

Line *l* bisects AB

bisection method
a method of solving an equation by successive approximation. Initially an interval is found which contains a root of the equation; the mid-point of the interval is then used as the next

approximation.

Example Solve $x^2 = 20$.

There is a root between 4 and 5. The mid-point, ie 4·5, is taken as the first approximation. Since $4·5^2 = 20·25$, this approximation is too big. So the root lies in the interval 4 to 4·5; the mid-point of this interval, 4·25 becomes the next approximation etc.

bisector
a straight line or plane that bisects an angle or a line segment – see MEDIAN

bivariate
of, involving, or containing two variables ⟨~ *distribution*⟩

Bombelli, Rafael (1526–73) Italian mathematician who did much work investigating the solution of cubic equation. This produced some of the earliest ideas about imaginary numbers.

Boole, George (1815–64) British mathematician who invented the branch of algebra which bears his name

Boolean
of or being a type of algebra in which logical symbols are used to represent relations between sets, and which is used extensively in the theory of computer programming ⟨~ *expression*⟩

bound
1 LOWER BOUND
2 UPPER BOUND

brace
SQUARE BRACKET

branch
a separate part of a mathematical curve ⟨*the two ~es of the hyperbola*⟩

breadth
distance from side to side; width, broadness

Briggs, Henry (1561–1631) Savilian Professor of Mathematics at Oxford who compiled the first set of common logarithm tables

Buffon, Comte de (1707–88) inventor of a famous experiment to estimate the value of π. In this, a needle of length l is dropped at random onto a set of equally spaced parallel lines. The probability that the line will intersect a line is $2l/\pi d$, where d is the distance between the lines.

by
– used in division to show that the preceding number is to be shared ⟨*divide 70 ~ 35*⟩, in multiplication ⟨*multiply 10 ~ 4*⟩, and in measurements ⟨*a room 15 feet ~ 20 feet*⟩

C

calculation
the process or an act of determining something (eg the answer to an arithmetical expression) by the principles and techniques of mathematics; *also* the result of such an act

calculator
1 a set or book of tables used in making calculations
2 (*also* **pocket calculator**) a small electronic device for performing arithmetical and other mathematical operations

calculus
1 a method of computation or calculation (eg formal logic) using a special symbolic notation
2 a branch of mathematics that deals with the nature of functions as infinitesimally small changes are made in their variables and with the ideas of limits, and that is basically composed of differential calculus and integral calculus – see DIFFERENTIAL CALCULUS, INTEGRAL CALCULUS

canonical *or* **canonic**
reduced to the simplest or clearest equivalent mathematical form ⟨*a ~ matrix*⟩

canonical form
the simplest form of a matrix; *specif* the form of a square matrix that has zero elements everywhere except along the principal diagonal (leading from upper left to lower right)

Cantor, Georg (1845–1918) German mathematician, noted for his work in analysis

cap
the mathematical symbol ∩ indicating an intersection (set containing the common elements of two sets) – compare CUP

> **Example** If A = {1, 2, 3, 4, 5} and B = {2, 4, 6, 8} then
> A ∩ B = {2, 4}

capacity
the measured ability to contain; volume ⟨*a jug with a ~ of 2 litres*⟩

Cardan, Geronimo (1501–76) Italian mathematician, author of *Ars Magna*, in which the general solution of a cubic equation was published for the first time

cardinal number *also* **cardinal**
1 a number (eg 4 or 15) that is used in the counting of whole

numbers and that indicates how many elements there are in a set – compare ORDINAL NUMBER

2 the property common to all mathematical sets having the same number of elements

cardioid

the heart-shaped curve traced by a point on the circumference of a circle that rolls completely round a fixed circle of equal radius. The equation of a cardioid in polar coordinates is $r = 1 + \cos \theta$.

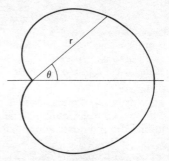

Carroll, Lewis. Pen-name of the Oxford mathematician Charles Lutwidge Dodgson (1832–98), author of *Alice in Wonderland*.

Cartesian coordinate

1 either of two numbers fixing a point's position on a plane (eg on a graph or map) by its distance from each of two straight axes drawn normally at right angles to each other – see ABSCISSA, ORDINATE; compare POLAR COORDINATE

2 any of three numbers fixing a point's position in space by its distance from each of three planes lying usu at right angles to each other – compare SPHERICAL COORDINATE

Cartesian plane

a plane whose points are assigned Cartesian coordinates

Cartesian product

a set that is constructed from two given sets and comprises all pairs of elements such that one element of the pair is from the first set and the other element is from the second set – called also DIRECT PRODUCT

catenary

the curve assumed by a perfectly flexible inextensible cord of uniform density and cross section hanging freely from two fixed points; the graph of $y = \cosh x$

Cauchy, Augustin-Louis (1789–1857) French mathematician, especially remembered for his work on functions of a complex variable

Cayley, Arthur (1821–95) British mathematician famous for his work on matrices and groups

Cayley table

a table representing the elements of a group and their products

Celsius
conforming to or being a scale of temperature on which water freezes at 0° and boils at 100° under standard conditions

centi-
one hundredth (10^{-2}) part of (a specified unit) ⟨centi*metre*⟩

centigrade
CELSIUS

centigram
one hundredth of a gram

centilitre
one hundredth of a litre

centimetre
one hundredth of a metre

centimetre-gram-second
of or being a system of units based on the centimetre as the unit of length, the gram as the unit of mass, and the second as the unit of time

central angle
an angle formed by two radii of a circle [an angle at the centre]

central limit theorem
a fundamental theorem in statistics which states that if means of various samples of a variable are plotted on a graph, the curve obtained will approximate closely to the normal distribution curve for that variable in the population as a whole, providing that the samples are sufficiently large

centre of curvature
the centre of the circle that has the same curvature as a curve at a given point

centre of gravity
the point in a rigid body where the weight of the body may be considered to act – compare CENTRE OF MASS

centre of mass
the point in a rigid body where the mass of the body may be considered to act. In a uniform gravitational field this is the same point as the centre of gravity – compare CENTROID

Examples
PLANE FIGURES:
Triangle   at the intersection of the medians.

Rectangle at the intersection of the diagonals.
Semi-circle on axis of symmetry, $4r/3\pi$ from the centre.
SOLIDS:
Cone on axis of symmetry, $\frac{1}{4}h$ from base.
Hemisphere on axis of symmetry $3r/8$ from base.
HOLLOW FIGURES:
Hemispherical shell on axis of symmetry, $\frac{1}{2}r$ from base.

centre of symmetry

the point about which a geometric figure displays symmetry

centroid

the geometrical centre of a
plane figure or solid object. For
a triangle, this is the point of
intersection of the medians. If
the figure or object has uniform
density the centroid is the same
as the centre of mass – compare
CENTRE OF GRAVITY, CENTRE
OF MASS

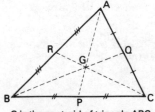

G is the centroid of triangle ABC

chain

1 a measuring instrument of 100 links used in surveying
2 a unit of length equal to 66ft (about 20·12m)

chain rule

a mathematical rule concerning the differentiation of a composite
function (eg f$[u(x)]$) by which under suitable conditions of
continuity and differentiability one function is differentiated with
respect to the second considered as an independent variable and
then the second function is differentiated with respect to the
independent variable: $\dfrac{\mathrm{d}y}{\mathrm{d}x} = \dfrac{\mathrm{d}y}{\mathrm{d}u} \times \dfrac{\mathrm{d}u}{\mathrm{d}x}$

Example Differentiate $y = (1 + \sin x)^5$.
Let $u = 1 + \sin x$,
then $y = u^5$,
$\dfrac{\mathrm{d}y}{\mathrm{d}u} = 5u^4$ and $\dfrac{\mathrm{d}u}{\mathrm{d}x} = \cos x$
$\dfrac{\mathrm{d}y}{\mathrm{d}x} = \dfrac{\mathrm{d}y}{\mathrm{d}u} \times \dfrac{\mathrm{d}u}{\mathrm{d}x}$
$= 5u^4 \times \cos x$
$= 5(1 + \sin x)^4 \cos x$

chance

the possibility of a specified or favourable outcome in an uncertain
situation (eg when a fair die is rolled the chance of scoring a six is
1 in 6) – see PROBABILITY

characteristic
the part of a common logarithm in front of the decimal point (eg the logarithm of 20 is 1·3010 and the characteristic is 1) – compare MANTISSA

check
(a sample or unit used for) the act of testing or verifying

chi-square
a statistical value that is a sum of terms each of which is a quotient obtained by dividing the square of the difference between the observed and expected values of a quantity by the expected value

chi-square distribution *also* **chi-squared distribution**
a frequency distribution that uses the properties of chi-square to test for statistical significance. The chi-square distribution can be used to determine how frequently a statistical distribution might arise by chance in an experiment.

chord
a straight line joining two points on a curve

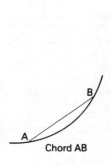

Chord AB

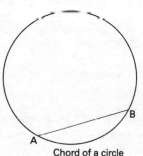

Chord of a circle

cipher (*in British usage also* **cypher**)
1 the arithmetical symbol 0; ZERO 1
2 any of the arabic numerals (0 to 9)

circle
1a a closed plane curve every point of which is equidistant from a fixed point within the curve; the locus of a point which moves in a plane so that its distance from a fixed point is constant
1b the plane surface bounded by such a curve
2 a circle formed on the surface of a sphere by the intersection of a plane ⟨~ *of latitude*⟩ – see also GREAT CIRCLE

circular
having the form of a circle

circular function
TRIGONOMETRIC FUNCTION

circular measure

circular measure
the measurement of angles in radians

circumcentre
– see CIRCUMCIRCLE

circumcircle
of a triangle the circle which
passes through the three
vertices of a triangle. Its centre
(called the *circumcentre*) is
found by bisecting the sides of
the triangle.

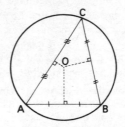

Circumcircle of triangle ABC

circumference
(the length of) the perimeter of a circle. The circumference of a
circle is equal to $2\pi r$, where r is the radius.

circumscribe
to draw or be drawn round in such a way that contact is made at as
many points as possible ⟨*a polygon* circumscribing *a circle*⟩

class
1 a group of adjacent values of a random variable
2 SET

class interval
CLASS 1; *also* the width of a statistical class

clockwise rotation
rotation in the same direction
as the hands of a clock. In
mathematics, angles measured
clockwise are given a negative
sign; anticlockwise angles are
given a positive sign.

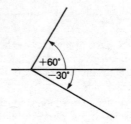

closed
1 *of a curve* traced by a moving point that returns to its starting
point without retracing its path; *also* so formed that every cross
section is a closed curve ⟨*a sphere is a* ∼ *surface*⟩
2 characterized by mathematical elements that when subjected to
an operation produce only elements of the same set (eg the set of
whole numbers is closed under addition and multiplication; you
cannot get a fraction by adding or multiplying them)

co-
of or being the angle which, with a given angle, makes 90°
⟨co*sine*⟩ ⟨co*tangent*⟩ – see also COFUNCTION

co-domain
a set that includes all the images of a given function – compare
DOMAIN, RANGE

>   **Example** A (possible) co-domain of the function
> $$x \mapsto x^2, \text{ where } x \in \mathbb{R}$$
>   is the set of real numbers. (The range, however, is the set of
>   real numbers greater than, or equal to, zero.)

coefficient
the number or mathematical quantity by which a variable is
multiplied ⟨*the* ~ *of* $3x^2$ *is 3*⟩

coefficient of correlation
CORRELATION COEFFICIENT

cofactor
the minor of an element of a square matrix or determinant,
assigned to be positive or negative according to its position in the
matrix or determinant

cofunction
a trigonometric function whose value for the complement of an
angle is equal to the value of a given trigonometric function for the
angle itself (eg the sine is the cofunction of the cosine)

collinear
lying on the same straight line
⟨~ *points*⟩

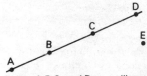

Points A,B,C, and D are collinear,
but A,B,C and E are not

combination
any of the different sets of a usu specified number of things that
can be chosen from a group and are considered *without regard to
order* within the set – compare PERMUTATION

>   **Example** The ten possible combinations of three letters selected
>   from *a, b, c, d, e,* are:
>   >   *abc, abd, abe, acd, ace, ade, bcd, bce, bde, cde.*
>   The number of combinations of *r* objects, selected from *n*
>   objects, is given by the formula
> $$\frac{n!}{r!(n-r)!}$$

This number is normally denoted by ^nC_r or $\binom{n}{r}$.

For the example above
$$^5C_3 = \frac{5!}{3!2!}$$
$$= \frac{5 \times 4}{1 \times 2}$$
$$= 10$$

combinatorial
of the manipulation of mathematical elements within sets that have a finite number of elements ⟨~ *mathematics*⟩

commensurable
having a common measure; *specif* divisible by a common unit a whole number of times

common
belonging equally to two or more quantities ⟨*triangles having a ~ side*⟩

common denominator
a common multiple of the denominators of several fractions

    Example The common denominator of the fractions $\frac{1}{3}$ and $\frac{1}{5}$ is 15.

$$\frac{1}{3} + \frac{1}{5} = \frac{5}{15} + \frac{3}{15} = \frac{8}{15}$$

common divisor
COMMON FACTOR

common factor
a number that divides two or more numbers without remainder (eg 6 is a common factor of 24 and 60; the *highest* common factor (HCF) is 12) – see also HIGHEST COMMON FACTOR; compare CO-PRIME

common logarithm
a logarithm whose base is 10 – often abbreviated to lg (eg lg 100 = 2)

common multiple
a multiple of each of two or more numbers (eg 60 is a common multiple of 6 and 15; the *lowest* common multiple (LCM) is 30) – see also LOWEST COMMON MULTIPLE

commutative
1 of or being a mathematical operation (eg addition) such that $x * y = y * x$, where $*$ denotes the operation. For example, multiplication is commutative ($3 \times 5 = 5 \times 3 = 15$); division is not ($2 \div 6 \neq 6 \div 2$)

2 having all elements obeying a commutative operation ⟨∼ *ring*⟩. A commutative group (also called an Abelian group) is one in which all elements can be associated in pairs that obey a commutative operation.

commute
of two mathematical quantities to give a commutative result

compasses (**a pair of**)
a geometrical instrument for drawing circles – compare DIVIDERS

complement
1 an angle or arc that when added to a given angle or arc equals 90°
2 the set of all things that do not belong to a given set (eg the complement of the set of even integers is the set of odd integers)

complementary angles
a pair of angles that have the sum of 90° (eg 20° and 70° are complementary angles)

completing the square
a way of solving quadratic equations by expressing the quadratic $ax^2 + bx + c$ in the form $a(x + p)^2 + q$

    Examples (a) $2x^2 + 12x + 25 = 2(x + 3)^2 + 7$
                 (b) $3x^2 + 6x + 1 = 3(x + 1)^2 - 2$

complex number
a number of the form $a + bi$ where a and b are real numbers and i is defined as the positive square root of minus one ($i = +\sqrt{-1}$) – see also ARGAND DIAGRAM, IMAGINARY NUMBER, REAL NUMBER

component
any of the vectors added together to form a given vector – compare RESOLVED PART

    Example If a vector **F** is inclined at 30° to the horizontal, its horizontal and vertical components are $F\cos30°$ and $F\sin30°$, respectively.

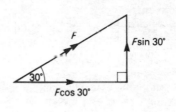

composite
having two or more factors; *esp, of a number* having two or more factors which are prime numbers; not prime ⟨*12, with factors of 2, 2, and 3, is a ∼ number*⟩

composite function
the function formed by applying two functions in succession.

    Example If f(x) = x^2 and g(x) = $x + 3$, the function gf(x) is the

result of applying function *f* and *then* function *g*, ie
$$gf(x) = x^2 + 3$$
Note, however, that fg(*x*) is different: fg(*x*) = $(x + 3)^2$

compound interest
interest computed on the original amount borrowed plus accumulated interest. If £*P* is invested at an annual rate of (compound) interest *r*% then after *n* years the value of the investment is given by
$$P\left(1 + \frac{r}{100}\right)^n$$

Example £1000 invested at 8% after 5 years is worth £1000 $(1 + 0.08)^5$ = £1469, correct to the nearest pound.

computation
1 computing, calculation
2 an amount computed

compute
to determine by mathematical means

concave
hollowed or rounded inwards like the inside of a bowl – compare CONVEX

concavity
1 a concave line or surface or the space included in it; a hollow
2 the quality of being concave

concentric
esp of circles and spheres having the same centre

conditional
1 true only for certain values of the variables or symbols involved ⟨~ *equations*⟩
2 stating the case when one or more random variables are fixed or one or more events are known ⟨~ *frequency distribution*⟩

conditional probability
the probability of an event which depends on the outcome of some other event

cone
1 the curved surface generated by rotating a line through 360° about another line (the axis), the angle between the lines being an acute angle (called the *semi-vertical angle* of the cone)
2 the solid object formed as above with an elliptical or circular

base. If the axis is perpendicular to the base the object is called a *right* cone.

A skew cone

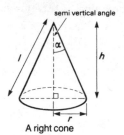

semi vertical angle

A right cone

Volume of a cone = $\frac{1}{3}\pi r^2 h$
Area of curved surface = $\pi r l$

confidence interval
a set of values within which there is a specified probability (eg 95 per cent) of including the true value of a statistical mean, average, variance, etc

confidence limits
the end points of a confidence interval

conformal
leaving relative sizes and angles unchanged after transformation

congruence *or* **congruency**
1 the quality or state of agreeing or coinciding; being congruent
2 a statement that two mathematical expressions are congruent with respect to a modulus

congruent
1 identical in size and shape $\langle \sim \textit{triangles} \rangle$ – compare SIMILAR

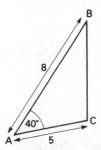

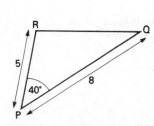

2 having the difference divisible by a specified number (the modulus) (eg 12 is congruent to 2 (modulo 5) since $12 - 2 = 2 \times 5$)

conic

conic *or* **conic section**
1 a plane curve, line, or point that is traced on the surface of a cone when an imaginary plane cuts through the cone

In the diagram
(a) plane perpendicular to the axis –
a CIRCLE,
(b) plane inclined to the axis, but not parallel to a generator – an ELLIPSE,
(c) plane parallel to a generator – a PARABOLA,
(d) plane parallel to the axis – a HYPERBOLA.

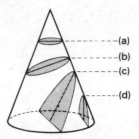

2 a mathematical curve generated by a point which moves so that the ratio of its distance from a fixed point (the focus) to its distance from a fixed line (the directrix) is constant
In the diagram PS/PM = *e*, the eccentricity.
For a parabola, $e = 1$
an ellipse, $e < 1$
a hyperbola, $e > 1$

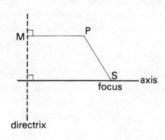

conjugate
1 CONJUGATE COMPLEX NUMBER
2 an element of a mathematical group that is equal to a given element of the group multiplied on the right by another element and on the left by the inverse of the latter element; if *a,b* are members of a group G, then $b^{-1}ab$ is a conjugate of *a*

conjugate complex number
either of two complex numbers (eg $a + bi$ and $a - bi$) which differ only in the sign connecting the two parts

consecutive
following one after the other in order without gaps (eg 10, 11, 12, 13 … are consecutive integers)

constant
a term in a mathematical expression that has a fixed value (eg in the polynomial $2x^3 + 5x^2 + 7x + 3$, the constant term is 3)

constant of integration
the constant term in an indefinite integral (which would become

zero on differentiation), eg in the indefinite integral
$$\int 3x^2\,dx = x^3 + c,$$
c is the constant of integration

construct
to draw (a geometrical figure) with suitable instruments, esp ruler and compasses, and under given conditions

contingency table
a table that tabulates the frequency distribution of one variable in the rows and that of another variable in the columns and that is used esp in the study of correlation between the variables

continuity
the property characteristic of a continuous mathematical function

continuous
having the property that the difference in the values of a mathematical function at two points may be made as small as possible by choosing two points sufficiently close together. Formal definition: the function $f(x)$ is continuous at $x = a$ if

$$f(a) = \lim_{x \to a} f(x).$$

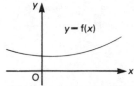

f(x) is a continuous function

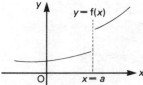

f(x) is discontinuous at $x = a$

convergent
having the nth term or the sum of the first n terms approach a finite limit as n increases without bound ⟨$a \sim sequence$⟩ ⟨$a \sim series$⟩

Example The series $1 + \frac{1}{2} + \frac{1}{4} + \frac{1}{8} + \ldots$ is convergent; its sum to infinity is 2

converse (of a statement)
the statement formed by interchanging the assumption and the result of another statement. (NB the statements may, or may not, be true.)

Example
Statement: 'All multiples of 8 are even numbers'. (True)
Converse: 'All even numbers are multiples of 8'. (False)

conversion
the process of changing one set of units into another (eg the rule

for converting Celsius temperatures into Fahrenheit is
F = ⅑C + 32

convex
curved or rounded outwards like the outside of a bowl – compare
CONCAVE

convexity
1 a convex line, surface, or part
2 the quality of being convex

convex polygon
a polygon in which all the interior angles are less than 180°. If at
least one angle is more than 180°, the polygon is concave.

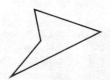

A convex pentagon   A concave quadrilateral

coordinate
any of a set of numbers used to specify the location of a point on a
line, in a plane, or in space – compare CARTESIAN COORDINATE,
POLAR COORDINATE

coordinate geometry
ANALYTICAL GEOMETRY

Copernicus, Nicholas (1473–1543) Polish astronomer who first
showed that the earth moves round the sun. He also made some
useful contributions to trigonometry.

coplanar
lying or acting in the same plane surface

co-prime
having no common factor (eg 6 and 35 are co-prime)

correlation
an association between two variables such that a change in one
implies a corresponding change in the other

correlation coefficient
a number or function that indicates the degree of correlation
between two sets of data or between two variables

correspondence
an association of one or more members of one set with one or
more members of another set

corresponding angles

angles that agree in position. In the diagram, the following pairs of angles are corresponding angles:

a and *p*, *b* and *q*, *c* and *r*, *d* and *s*.

If the lines l_1 and l_2 are parallel, the corresponding angles are equal.

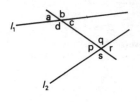

cosecant

the mathematical function that is the reciprocal of sine

coset

a subset of a mathematical group that consists of all the products obtained by multiplying, either on the right or on the left, a fixed element of the group by each of the elements of a given subgroup

cosine

a trigonometric function of an angle. The cosine of an acute angle θ is the ratio of the side adjacent to the angle to the hypotenuse in a right-angled triangle;

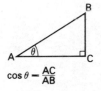

$$\cos\theta = \frac{AC}{AB}$$

$$\cos\theta = \frac{\text{adjacent side}}{\text{hypotenuse}}$$

The cosine of a general angle θ is the ratio *x/r* in triangle OPM, in which point P has coordinates (x, y) and *r* is the length OP. The coordinates *x* and *y* obey the usual sign convention; the length *r* is always positive – compare SINE, TANGENT

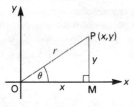

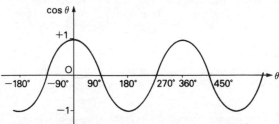

Graph of the function $\theta \rightarrow \cos\theta$

cosine rule

In a triangle ABC,
$$a^2 = b^2 + c^2 - 2bc \cos A$$
– compare SINE RULE

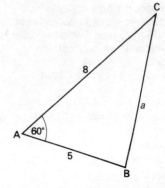

Example Given that $b = 8$, $c = 5$ and $A = 60°$, find a.

$$a^2 = 8^2 + 5^2 - 2 \times 8 \times 5 \times \cos 60°$$
$$= 64 + 25 - 40$$
$$= 49$$
$$a = 7$$

cotangent

the mathematical function that is the reciprocal of tangent

count

to indicate or name by units or groups so as to find the total number of units involved

countable

of a set capable of having its members in a one-to-one correspondence with the natural numbers 1, 2, 3, 4, ...

Example the set $\{1, 4, 9, 16, ...\}$ is countable.

$$
\begin{array}{cccc}
1 & 2 & 3 & 4 & ... \\
\updownarrow & \updownarrow & \updownarrow & \updownarrow & \\
1 & 4 & 9 & 16 & ...
\end{array}
$$

counter-example

an example employed to show that a proposition is false

Example Consider the proposition:
"All prime numbers are odd".
This is shown to be false by the counter-example:
"2 is an *even* prime number."

covariance

1 the expected value of the product of the differences of two random variables from their respective means
2 the arithmetic mean of the products of the differences of corresponding values of two random variables from their respective means

Cramer, Gabriel (1704–52) German mathematician, chiefly remembered for Cramer's rule, a method for solving linear equations

cross product

VECTOR PRODUCT

cube

1 a three-dimensional geometric shape having six equal square faces

2 the number resulting from multiplying a number by itself twice (eg $5^3 = 5 \times 5 \times 5 = 125$ is the cube of 5)

Formula for the sum of the first n cubes:
$$1^3 + 2^3 + 3^3 + \ldots + n^3 = \tfrac{1}{4}n^2(n + 1)^2$$

cube root

a number that when cubed produces a particular number ⟨*3 is the ~ of 27*⟩

cubic curve

a curve obtained by plotting $y = ax^3 + bx^2 + cx + d$

cubic equation

an equation of the form $ax^3 + bx^2 + cx + d = 0$

cubic measure

a unit (eg cubic metre or cubic inch) for measuring volume

cubic polynomial

a polynomial of the form $ax^3 + bx^2 + cx + d$

cubit

any of various ancient units of length based on the length of the forearm from the elbow to the tip of the middle finger and usu equal to about 0·46m (about 18in)

cuboid

a three-dimensional geometric shape having six rectangular faces

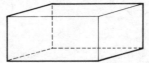

cumulative

including or considering together all values less than or less than and equal to a particular value ⟨*~ distribution*⟩ ⟨*~ frequency*⟩

Example (of cumulative frequency) The number of children in a family

No. of children per family	Frequency, ie no. of families	Cumulative frequency
0	5	5
1	6	11
2	8	19
3	4	23
4	2	25
5	0	25

cup

the mathematical symbol ∪ indicating a union (set containing all the elements of two sets) – compare CAP

Example If A = {1, 2, 3, 4, 5} and B = {2, 4, 6, 8} then
A∪B = {1, 2, 3, 4, 5, 6, 7, 8}

curvature
the rate of change of the angle through which the tangent to a
curve turns in moving along the curve; $d\psi/ds$ where ψ is the angle
at which the tangent is inclined to the horizontal and s is the arc
length

Formula: curvature, $\varkappa = \dfrac{\dfrac{d^2y}{dx^2}}{\left\{1 + \left(\dfrac{dy}{dx}\right)^2\right\}^{3/2}}$

NB Radius of curvature = $1/\varkappa$

curve
1 a representation on a graph of a varying quantity (eg speed
varying with time)
2 a continuous set of points, in a plane, whose coordinates are
defined by an equation, or by geometrical conditions – see LOCUS

curvilinear
consisting of or bounded by curved lines; represented by a curved
line

cusp
a fixed point on a mathematical
curve at which a point tracing
the curve would exactly reverse
its direction of motion and
begin to trace a mirror image of
the curve

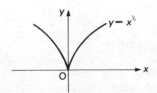

The graph of $y = x^{2/3}$ has a cusp at (0,0)

cycle
a permutation of a set of ordered mathematical elements (eg
numbers) in which each element takes the place of the next and
the last becomes first

cyclic group
a mathematical group that contains an element from which every
other element of the group can be derived by repeatedly applying
an operation (eg addition) which is defined for the group

cyclic quadrilateral

a quadrilateral whose vertices lie on the circumference of a circle.

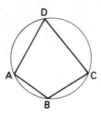

The opposite angles of a cyclic quadrilateral are supplementary; in the diagram, $\hat{A} + \hat{C} = 180°$ and $\hat{B} + \hat{D} = 180°$.

cycloid

the curved path traced out by a point on the circumference of a circle rolling along a straight line

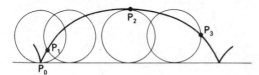

cylinder

1 a surface formed by rotating a line through 360° about a parallel line

2 a solid whose curved surface is cylindrical and whose ends are circular

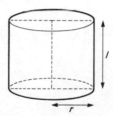

Volume $= \pi r^2 l$

Area of curved surface $= 2\pi r l$

cylindrical coordinate

any of the three coordinates r, θ, or z used to define the location of a point in space where, in one of three planes, r is the distance on a line drawn from a fixed point to the point being defined, θ is the angle that this line makes with a fixed line originating from the fixed point, and z is the perpendicular distance of the point being defined from the fixed line

D

d'Alembert, Jean le Rond (1717–83) French mathematician, named after the church of St Jean le Rond in Paris, where he was found after being abandoned as a child; famous for his work on the roots of equations in complex numbers and on the solution of partial differential equations

data
factual information (eg measurements or statistics) used as a basis for reasoning, discussion, or calculation; *esp* the information given in a mathematical exercise

datum
something (eg a number, fixed point, or assumed value) used as a basis for calculating or measuring

dec- *or* **deca-** *also* **dek-** *or* **deka-**
ten ⟨*decathlon*⟩

decade
a period of 10 years

decagon
a 10-sided polygon

decahedron
a solid with 10 faces

deci-
one tenth (10^{-1}) part of (a specified unit) ⟨deci*litre*⟩

decigram
one tenth of a gram

decile
any of nine values in a frequency distribution that divide it into 10 parts (intervals) each containing $\frac{1}{10}$ of the individuals, items, etc under consideration; *also* any of the 10 groups of individuals, items, etc comprising such an interval

decilitre
one tenth of a litre

decimal (*adjective*)
1 based on the number 10; *specif* subdivided into units which are tenths, hundredths, etc of another unit
2 expressed as a decimal fraction

decimal (*noun*) *or* **decimal fraction**
a proper fraction (fraction having a value of less than one) that is

expressed as a sum of multiples of powers of $\frac{1}{10}$ by writing a dot followed by one digit for the number of tenths, one digit for the number of hundredths, and so on (eg $0 \cdot 25 = \frac{25}{100}$)

decimalize *or* **decimalise**
to convert to a decimal system

decimetre
one tenth of a metre

decrement
a negative mathematical increment

Dedekind, Julius (1831–1916) German mathematician who placed the theory of numbers (especially irrational numbers) on a logical basis

deduct
to take away (an amount) from a total; subtract

definite integral
the difference between the values of an integral at two given limits – compare INDEFINITE INTEGRAL, INTEGRAL

$$\textbf{Example} \int_2^4 3x^2\mathrm{d}x = \left[x^3\right]_2^4$$
$$= 4^3 - 2^3$$
$$= 56$$

degree
1 (*symbol* °) a unit of measurement for angles, 360° equalling one rotation – compare RADIAN
2 a unit for measuring temperature
3 the highest power of the variable present in a polynomial (eg the polynomial $5x^4 + 7x^2 - 3x + 2$ is of degree 4)
4 the sum of the exponents of an expression which is the product of several variables (eg x^2yz^3 has a degree of 6)

degree of accuracy
– see ACCURACY

degree of freedom
the number of independent values or quantities which must be specified to completely define a statistical situation

delta
an increment of a variable. For example, δx means 'a small increase in the value of x'.

demography
the statistical study of human populations, esp with reference to size, density, and distribution

de Moivre, Abraham (1667–1754) a French-born mathematician who fled to England, where he worked with Halley and Newton

de Moivre's theorem
a theorem of complex numbers:
$(\cos x + i \sin x)^n = (\cos nx + i \sin nx)$

de Morgan, Augustus (1806–71) Professor of Mathematics at London University

de Morgan's laws
laws governing the intersection and union of the complements of two sets:
$$(A \cup B)' = A' \cap B'$$
$$(A \cap B)' = A' \cup B'$$

denary
of, being, or belonging to a system of numbers having 10 as its base; decimal

denominator
the part of a fraction that is below the line and that indicates how many parts the numerator is divided into (eg the denominator of the fraction ⅗ is 5) – see COMMON DENOMINATOR

density
the mass of a particular substance per unit volume, ie density = mass/volume. Density is sometimes used for mass per unit area or mass per unit length. Population density refers to the number of individuals per unit area.

dependent variable
a variable whose value is determined by that of one or more independent variables in a function (eg in $z = x^2 + 3xy + y^2$, z is the dependent variable) – compare INDEPENDENT VARIABLE, DUMMY VARIABLE

depression
– see ANGLE OF DEPRESSION

depth
the perpendicular measurement downwards from a surface

derivative
of a function f(x) when x = a the value of the derived function $f'(x)$ when $x = a$, ie $f'(a)$. (eg $f(x) = x^3$, $f'(x) = 3x^2$. The derivative, when $x = 2$, is 12.) However, derivative is often used to mean derived function.

derived function ($f'(x)$)
of a function f(x) is

$$f'(x) = \lim_{h \to 0} \frac{f(x + h) - f(x)}{h}$$

It is sometimes called the gradient function because $f'(a)$ is the gradient of the curve $y = f(x)$ when $x = a$ – see also
DIFFERENTIAL COEFFICIENT

Examples

$f(x)$	$f'(x)$
x^n	nx^{n-1}
$\sin x$	$\cos x$
$\cos x$	$-\sin x$
$\tan x$	$\sec^2 x$
e^x	e^x
$\ln x$	$1/x$

derived unit
a unit (eg the newton, pascal, or watt) defined in terms of the basic units of a system (eg the SI system)

Descartes, René (1596–1650) French mathematician whose application of algebraic methods (using Cartesian coordinates) laid the foundation of much of the new developments of mathematics in the following centuries

determinant
a square array of numbers, bordered on either side by a vertical line, whose value is calculated by adding and multiplying the numbers according to a complex rule, and that is used in the study of matrices and to solve simultaneous equations

$$\begin{vmatrix} a_1 & a_2 \\ b_1 & b_2 \end{vmatrix} = a_1 b_2 - a_2 b_1$$

Example

$$\begin{vmatrix} 3 & 2 \\ 4 & 5 \end{vmatrix} = 3 \times 5 - 2 \times 4 = 7$$

$$\begin{vmatrix} a_1 & a_2 & a_3 \\ b_1 & b_2 & b_3 \\ c_1 & c_2 & c_3 \end{vmatrix} = a_1(b_2 c_3 - b_3 c_2) - a_2(b_1 c_3 - b_3 c_1) + a_3(b_1 c_2 - b_2 c_1)$$

Example

$$\begin{vmatrix} 3 & 2 & 5 \\ 2 & 3 & 0 \\ 1 & 1 & 2 \end{vmatrix} = 3 \times (3 \times 2 - 0 \times 1) - 2 \times (2 \times 2 - 0 \times 1) + 5 \times (2 \times 1 - 3 \times 1)$$
$$= 3 \times 6 - 2 \times 4 + 5 \times (-1)$$
$$= 5$$

The determinant of a (square) matrix, det (**M**), is formed by treating the matrix **M** as a determinant.

Example

$$\mathbf{M} = \begin{pmatrix} 1 & 3 \\ 2 & 9 \end{pmatrix} \quad \det(\mathbf{M}) = \begin{vmatrix} 1 & 3 \\ 2 & 9 \end{vmatrix} = 9 - 6 = 3$$

determine
to fix the form or character of beforehand ⟨*two points ~ a straight line*⟩

deviation
the difference between a single value in a series of numbers and the mean of those numbers, ie if $\bar{x}$ is the mean of $x_1, x_2, x_3, \ldots x_n$, the deviation of x_i is $x_i - \bar{x}$

di-
twice; twofold; double

diagonal (*adjective*)
joining two nonadjacent angles or corners of a geometric figure

diagonal (*noun*)
a diagonal straight line

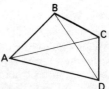

AC and BD are the diagonals of quadrilateral ABCD

diagonal matrix
a square matrix in which all the elements are zero except those on the leading diagonal (top left to bottom right)

Example
$$\begin{pmatrix} 1 & 0 & 0 \\ 0 & 3 & 0 \\ 0 & 0 & 4 \end{pmatrix}$$

diameter
1 a line passing through the centre of a geometric figure or body

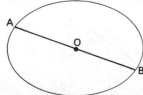

AB is a diameter of the circle   AB is a diameter of the ellipse

2 the length of a straight line from one side of an object, esp a circle, to the other, passing through its centre – compare RADIUS

3 *of a parabola* the locus of the midpoints of chords of the parabola

difference

1 the degree or amount by which things differ in quantity or measure; *specif* the result of subtracting one number from another

2 *of two sets* the (symmetric) difference is the set represented by the shaded part of the Venn diagram

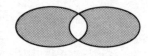

3 *of two squares*
$$a^2 - b^2 = (a + b)(a - b)$$

difference equation

a mathematical equation which defines successive members of a sequence in terms of sums of, or differences between, other members of the sequence

> **Example** $u_n = u_{n-1} + u_{n-2}$, where $u_1 = 1$ and $u_2 = 1$.
> (This particular difference equation defines the Fibonacci sequence 1, 1, 2, 3, 5, 8, 13, 21, ...)

differentiable

a function $f(x)$ is said to be differentiable when $x = a$ if its derived function exists when $x = a$. To be differentiable $f(x)$ and $f'(x)$ must both be continuous when $x = a$.

> **Examples** (a) $f(x) = x^2$ is differentiable for all values of x.
>
> (b) $f(x) = x^{2/3}$ (and hence $f'(x) = \frac{2}{3}x^{-1/3}$)
>
> is differentiable for all values of x except $x = 0$.

differential

1 the product of the derivative of a function of one variable with the increment of the independent variable.

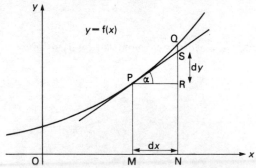

In the diagram, PS is the tangent to the curve at P, so
$\tan\alpha = f'(x)$, and MN = dx is an arbitrary (small) increase in x.

$$\text{Hence } dy = RS$$
$$= \tan\alpha \ dx$$
$$= f'(x) \ dx.$$

dy and dx are called differentials.

2 TOTAL DIFFERENTIAL

differential calculus
a branch of mathematics discovered simultaneously by Newton
and Leibnitz that deals chiefly with the rate of change of functions
with respect to their variables and is used to find the slope of a
curve at any point and derivatives of functions

differential coefficient (dy/dx)
– see DERIVED FUNCTION, DERIVATIVE
If $y = f(x)$
then $dy/dx = f'(x)$
(eg $y = x^3$, $dy/dx = 3x^2$)

differential equation
an equation containing differential coefficients or derivatives of
functions

    Examples

(a) $\dfrac{dy}{dx} = 3x^2$

(b) $\dfrac{dy}{dx} + 5y = x$

(c) $\dfrac{d^2y}{dx^2} - 5\dfrac{dy}{dx} + 6y = 12$

differentiate
to obtain the mathematical derivative of

differentiation
the act or process of differentiating

digit
1 any of the arabic numerals (0 to 9)
2 any of the elements that combine to form numbers in a system
other than the decimal system

digital
1 of calculation by numerical methods which use separate units
2 of data in the form of numerical units
3 *of an automatic device* presenting information in the form of
numerical digits rather than by a pointer moving round a dial
⟨~ *clock*⟩
4 *of a computer* operating with numbers expressed as separate
pulses representing digits

dihedral angle

dihedral angle *also* **dihedral**
the angle formed by two
intersecting planes; the angle
between two lines, one in each
plane, perpendicular to the
common line of the two planes

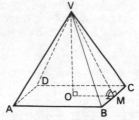

In the pyramid VABCD, the dihedral
angle between the inclined plane
VBC and the base ABCD is ∠VMO

dimension

1 measure in one direction; *specif* any of three or four coordinates
determining a position in space or space and time
2 any of a set of parameters necessary and sufficient to determine
uniquely each element of a system of usu mathematical entities
⟨*the surface of a sphere has two ~s*⟩
3 the number of elements in a basis of a vector space
4 any of the fundamental quantities, specif mass (M), length (L),
and time (T), in terms of which any physical quantity may be
expressed

Examples

Quantity	Dimensions
Area	L^2
Volume	L^3
Speed	LT^{-1}
Acceleration	LT^{-2}
Force	MLT^{-2}
Momentum	MLT^{-1}
Energy	ML^2T^{-2}
Density	ML^{-3}
Angle	dimensionless
Angular velocity	T^{-1}

Diophantine equation

an indeterminate polynomial equation with integral coefficients for
which it is required to find all integral solutions ⟨*x = 3, y = 4,
z = 5 is one solution of the* ~ $x^2 + y^2 = z^2$⟩ [named after
Diophantus of Alexandria, 3rd century AD]

directed

having a direction ⟨~ *line segment*⟩

directed number

a signed number (eg ... −3, −2, −1, 0, +1, +2, +3 ...)

direction cosine

the cosine of an angle between a vector and an axis

Example

In the diagram the direction cosines of $\overrightarrow{OA}$ are
$$\cos\alpha = 3/5, \qquad \cos\beta = 4/5$$

If **v** is the (three dimensional) vector
v = a**i** + b**j** + c**k**, and
$v = \sqrt{(a^2 + b^2 + c^2)}$
then the direction cosines of **v** are
$\cos\alpha = a/v$, $\cos\beta = b/v$,
$\cos\gamma = c/v$

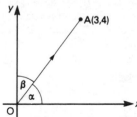

direct proportion

the relationship between two variables whose ratio is constant

Example If pencils are sold for 10 pence each, and the total cost of n pencils is C pence, then C is proportional to n, ie
$$C = 10n.$$

The expression 'y is proportional to x' is written $y \propto x$,
If $y \propto x$, then y is related to x by an equation of the form
$y = Kx$
If $y \propto x^n$, then y is related to x^n by an equation of the form
$y = Kx^n$ – compare INVERSE PROPORTION

directrix (*plural* **directrixes** *also* **directrices**)

a straight line that together with a fixed point (focus) forms the reference system for generating a conic section (eg an ellipse or parabola) in plane geometry

discontinuity

a point in the domain of a mathematical function at which it is not continuous (eg f(x) = $1/x$ has a discontinuity when $x = 0$)

discontinuous

of a mathematical function having one or more discontinuities

Example f(x) = $1/(x^2 - 4)$ is discontinuous when $x = +2$ and $x = -2$

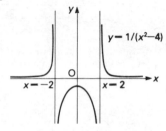

discrete

taking on or having a finite or countable number of values $\langle a \sim$ *random variable*$\rangle$

43

discriminant

discriminant
the discriminant of the quadratic equation $ax^2 + bx + c = 0$ is $b^2 - 4ac$.
If $b^2 - 4ac > 0$, the equation has two distinct real roots,
  $b^2 - 4ac = 0$, the equation has two identical real roots,
  $b^2 - 4ac < 0$, the equation has two complex roots

disjoint
having no elements in common
  Example

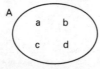

A = {a, b, c, d} B = {p, q, r, s} are disjoint sets

dispersion
the scattering or the extent of the scattering of the values of a frequency distribution from a mean

displacement
a vector representing the translation from one point to another point.
In the diagram, A is the point (1,2) and B (4,6). The displacement $\overrightarrow{AB}$ is given by $\binom{3}{4}$.

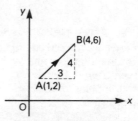

distance
the degree or amount of separation between two points, lines, surfaces, or objects measured along the shortest path joining them. In the Cartesian plane, the distance between (x_1, y_1) and (x_2, y_2) is given by the formula

$$\sqrt{(x_2 - x_1)^2 + (y_2 - y_1)^2}.$$

Example In the diagram for the displacement $\overrightarrow{AB}$ (see DISPLACEMENT), the distance between A and B is

$$\sqrt{(4 - 1)^2 + (5 - 1)^2}$$
$$= \sqrt{3^2 + 4^2}$$
$$= 5$$

distribution
1 FREQUENCY DISTRIBUTION

2 PROBABILITY FUNCTION
3 PROBABILITY DENSITY FUNCTION

distributive
of or being a mathematical law which relates the operations of addition and multiplication such that $x(y + z) = xy + xz$
(eg $5 \times (2 + 7) = 5 \times 9 = 45$,
$5 \times 2 + 5 \times 7 = 10 + 35 = 45$)

divergent
of a mathematical series having a sum that does not converge to a limit as the number of terms in the series increases (eg $1 + \frac{1}{2} + \frac{1}{3} + \frac{1}{4} + \ldots$ is a divergent series) – compare CONVERGENT

divide
1 to determine how many times (a number or quantity) contains another number or quantity by means of a mathematical operation $\langle \sim 42\ by\ 14 \rangle$
2 *of a number or quantity* to be contained in (another number or quantity) a whole number of times $\langle 14 \sim s\ 42 \rangle$

dividend
a number to be divided by another – compare DIVISOR, QUOTIENT
   Example In $15 \div 3 = 5$, the dividend is 15. (3 is the divisor and 5 the quotient.)

dividers
an instrument like a pair of compasses for measuring or marking (eg in transferring dimensions) that consists of two pointed arms jointed together

divisible
capable of being divided exactly $\langle 21\ is \sim by\ 3 \rangle$

division
an act, process, or instance of dividing one number by another

divisor
1 the number by which another number is to be divided – compare DIVIDEND, QUOTIENT
2 an integer (other than 1) which is a factor of another (eg 7 is a divisor of 21)

dodeca- *or* **dodec-**
twelve

dodecagon
a 12-sided polygon

dodecahedron (*plural* **dodecahedrons** *or* **dodecahedra**)
a solid with 12 faces. A regular dodecahedron is formed from 12 regular pentagons.

domain

the set of values to which a mathematical variable is limited; *esp* the set of values that the independent variable of a function may take – compare RANGE, CO-DOMAIN

> **Example** $f(x) = x^2$. If the domain is the set $\{1, 2, 3, 4, 5\}$ then the range is $\{1, 4, 9, 16, 25\}$.

> For many (real) functions, the domain is taken to be the set of real numbers, but in some cases it is necessary to restrict the domain, in order to define the function (eg the domain of $f(x) = \sqrt{x}$ is the set of non-negative real numbers).

dot

a point placed between two values as a multiplication sign

dot product

SCALAR PRODUCT

double

a multiply by two

double angle formulae

formulae used in trigonometry to find the sine, cosine, or tangent of an angle which is double a given angle:

$$\sin 2A = 2\sin A \cos A$$
$$\cos 2A = \cos^2 A - \sin^2 A$$
$$= 2\cos^2 A - 1$$
$$= 1 - 2\sin^2 A$$
$$\tan 2A = \frac{2\tan A}{1 - \tan^2 A}$$

– compare ADDITION FORMULAE

dummy variable

a mathematical variable chosen at random whose choice does not affect the meaning of the expression in which it occurs ⟨*the variable of integration in a definite integral is a* ∼⟩ – compare DEPENDENT VARIABLE, INDEPENDENT VARIABLE

> **Example** $\int_1^2 3x^2 \mathrm{d}x = \left[x^3\right]_1^2 = 2^3 - 1^3 = 7$

> $\int_1^2 3t^2 \mathrm{d}t = \left[t^3\right]_1^2 = 2^3 - 1^3 = 7$

> x and t are dummy variables

duo-

two

duodecimal

of, being, or belonging to a system of numbers having 12 as its base

Example The duodecimal number 234_{12} equals the decimal number $2 \times 12^2 + 3 \times 12 + 4 = 328$

dynamics

a branch of mechanics concerned with motion and the forces which cause motion (but not forces in equilibrium, or acting on a body which is at rest) – compare STATICS

E

e
1 (*in roman type*: e) the base of natural logarithms
$$e = \lim_{n \to \infty} (1 + 1/n)^n$$
e = 2·71828 correct to six significant figures
2 (*in italic type: e*) the symbol for eccentricity
3 (*in italic type: e*) the identity element of a group

eccentric
1 not having the same centre ⟨~ *spheres*⟩
2 deviating from a circular path ⟨*an* ~ *orbit*⟩
3 located elsewhere than at the geometrical centre; also having the axis or support so located ⟨*an* ~ *wheel*⟩

eccentricity (*symbol e*)
a number that for a given conic section is the ratio of the distances from any point on the curve to the focus (fixed point) and the directrix (specified straight line)
For   a parabola, $e = 1$,
      an ellipse, $e < 1$,
      a hyperbola, $e > 1$

edge
the common line of two adjoining faces of a polyhedron – see also EULER'S RELATION

eigenvalue
– see EIGENVECTOR

eigenvector
of a matrix M a column vector **X** such that $\mathbf{MX} = \lambda\mathbf{X}$, where λ is a real number (called the *eigenvalue*)

Example Consider $\mathbf{M} = \begin{pmatrix} 1 & -1 \\ 2 & 4 \end{pmatrix}$

A solution of the matrix equation $\mathbf{MX} = \lambda\mathbf{X}$ is

$\lambda = 2$ and $\mathbf{X} = \begin{pmatrix} 1 \\ -1 \end{pmatrix}$; another solution is

$\lambda = 3$ and $\mathbf{X} = \begin{pmatrix} 1 \\ -2 \end{pmatrix}$, ie $\begin{pmatrix} 1 & -1 \\ 2 & 4 \end{pmatrix} \begin{pmatrix} 1 \\ -1 \end{pmatrix} = \begin{pmatrix} 2 \\ -2 \end{pmatrix} = 2 \begin{pmatrix} 1 \\ -1 \end{pmatrix}$ and

$\begin{pmatrix} 1 & -1 \\ 2 & 4 \end{pmatrix} \begin{pmatrix} 1 \\ -2 \end{pmatrix} = \begin{pmatrix} 3 \\ -6 \end{pmatrix} = 3 \begin{pmatrix} 1 \\ -2 \end{pmatrix}$.

Thus 2 is an eigenvalue and $\begin{pmatrix} 1 \\ -1 \end{pmatrix}$ a corresponding

eigenvector, and 3 is another eigenvalue with eigenvector $\begin{pmatrix} 1 \\ -2 \end{pmatrix}$.

Einstein, Albert (1879–1955) Inventor of the Theory of Relativity

element
1 any of the members of a set. The symbol ∈ means "is an element of".
2 any of the numbers or symbols in a mathematical array (eg a matrix)
3 *esp in integration* a small component part of a geometrical figure (eg the elements of a disc are concentric rings, the elements of a cone are circular discs perpendicular to the axis)

elementary matrix
a matrix obtained by applying an elementary operation to an identity matrix

elementary operation
the elementary operations that can be performed on a matrix are:
  (a) interchanging two rows (or columns)
  (b) multiplying a row (or column) by a constant
  (c) adding a multiple of one row (or column) to another

Elements
a famous collection of geometrical proofs compiled by Euclid

elevation
1 a scale drawing of the front or side view of a three-dimensional object
2 – see ANGLE OF ELEVATION

eliminate
to cause (a quantity) to not appear in an equation by combining two or more equations

ellipse
the locus of a point P which moves so that its distance from a fixed

point S (the focus) is in a constant ratio *e* (the eccentricity) to its distance from a fixed line *l* (the directrix),

$$\frac{PS}{PM} = e, \text{ where } e < 1.$$

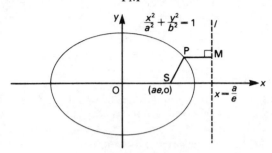

The equation of the ellipse referred to axes through O is

$$\frac{x^2}{a^2} + \frac{y^2}{b^2} = 1, \text{ (where } b^2 = a^2(1 - e^2))$$

An ellipse can alternatively be thought of as a two-dimensional closed curve generated by a point moving such that the sum of its distances from two foci is constant; the intersection of a plane cutting obliquely through a cone – compare CONIC SECTION, HYPERBOLA, PARABOLA

ellipsoid
1 a closed solid or surface obtained by rotating an ellipse about one of its axes
2 something shaped roughly like an ellipsoid; a somewhat flattened sphere (eg the earth) – compare HYPERBOLOID, PARABOLOID

empty
of a mathematical set having no elements (eg the solution set of $x + 1 = x$ is an empty set)

enlargement
a transformation of a geometrical figure in which all the linear dimensions are increased by the same factor (the scale factor). The angles are not changed by an enlargement – compare REFLECTION, ROTATION, SHEAR, TRANSLATION

enumerable
COUNTABLE

enumerate
to ascertain the number of; count

envelope
a curve that touches each of a
group of related curves at one
point only

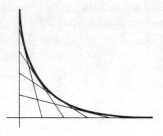

epicycloid
a curve traced by a point on the circumference of a circle that rolls
on the outside of a fixed circle – compare HYPOCYCLOID

equal
of the same quantity, amount, or number as another

equals sign *also* **equal sign, equality sign**
a sign = indicating mathematical or logical equivalence

equate
to make or set equal; equalize

equation
a mathematical statement that two expressions (containing one or
more 'unknowns', usu represented by x, y, z etc) are equal.
Finding the value(s) of the unknown(s) is called 'solving the
equation'; individual values are called roots of the equation.

Examples

Equation	Solution
$x + 2 = 5$	$x = 3$
$x + 1 = x + 3$	none
$x^2 - 5x + 6 = 0$	$x = 2$ or 3
$x^2 + 1 = 0$	none
$x - y = 0$	infinitely many solutions (eg $x = 7$, $y = 7$)

equiangular
having all or corresponding angles equal $\langle an \sim triangle \rangle$
$\langle \sim polygons \rangle$

equidistant
equally distant from two or more places

equilateral
having all sides equal $\langle \sim triangle \rangle$

equilibrant
a force or system of forces capable of balancing another force or
system of forces to produce equilibrium

equiprobable
having the same degree of logical or mathematical probability

equivalence *or* **equivalency**
the relation between two statements which are either both true or both false such that each implies the other

equivalent fraction
a fraction obtained by multiplying the top and the bottom of a given fraction by the same number

$$(\text{eg } \frac{2}{3} = \frac{2 \times 5}{3 \times 5} = \frac{10}{15}).$$

NB equivalent fractions are equal.

erect
to draw or construct (eg a perpendicular) on a given base

error
the difference between an approximation and its exact value

estimate
(the numerical value of) a rough or approximate calculation

Euclid Greek mathematician, 3rd century BC. Compiler of the Elements, a systematic collection of geometrical proofs.

euclidean *also* **euclidian** (*often cap E*)
of or being the geometry of Euclid or a geometry based on Euclid's assumptions regarding space; *specif* being a geometry in which there is only one line through a point which is parallel to some other line – compare HYPERBOLIC

euclidean space (*often cap E*)
three-dimensional space in which euclidean geometry applies

Euler, Leonhard (1707–83) Swiss mathematician who provided a systematic exposition of the work of Newton and Leibnitz

Euler's relation
a formula relating the number of faces, vertices, and edges of a polyhedron: $F + V = E + 2$, where F is the number of faces, V is the number of vertices, and E is the number of edges

Example For a cube, $F = 6$, $V = 8$, and $E = 12$
$$6 + 8 = 12 + 2$$

even
exactly divisible by two (eg $-4, -2, 0, 2, 4, 6 \ldots$ are even numbers)

even function
a mathematical function that does not change in value when the independent variable changes its sign, ie $f(a) = f(-a)$

Examples
(a) $f(x) = x^2, x^4, x^6, ...$
(b) $f(x) = \cos x$

– compare ODD FUNCTION

event
any of the possible outcomes of an experiment ⟨*7 is an* ~ *in the throwing of two dice*⟩

evolute
a curve that is made up of the collection of points that are the centres of curvature of some given curve (the involute) – compare INVOLUTE

exact differential
a mathematical expression that is the sum of the products of the partial derivatives of a function of several variables and the increment in their respective variables

example
1 a particular single item, fact, incident, or aspect that is representative of all of a group or type to which it belongs
2 an instance (eg a mathematical problem to be solved) serving to illustrate a rule or to act as an exercise in the application of a rule

exercise
something performed or practised in order to develop, improve, or display a specific power or skill ⟨*arithmetic* ~s⟩

expand
to express in fuller mathematical form, esp as the sum of many terms of a series

expansion
the result of expanding a mathematical expression or function – see BINOMIAL THEOREM

Examples
(a) $(1 + x)^4 = 1 + 4x + 6x^2 + 4x^3 + x^4$
(b) $(1 - x)^{-1} = 1 + x + x^2 + x^3 + ...$ (provided $-1<x<1$)

expected value
the mean value of a random variable

explicit function
a mathematical function defined by an expression containing only independent variables (eg in the expression $y = 3x^2 + 2x + 1$, y is an explicit function of x) – compare IMPLICIT FUNCTION

exponent
a symbol written above and to the right of a mathematical expression to indicate the number of times that a quantity is

multiplied by itself (eg in the expression a^3, the exponent 3 indicates that a is cubed)

exponential function *or* **exponential**
a mathematical function (eg $f(x) = 2^x$) in which an independent variable appears in an exponent; *specif* e^x

exponential series

$$1 + \frac{x}{1!} + \frac{x^2}{2!} + \frac{x^3}{3!} + \frac{x^4}{4!} + \ldots$$

the exponential series converges to e^x for all (real) values of x

exponentiation
the action or process of multiplying a quantity by itself the number of times specified by an exponent

expression
a mathematical or logical symbol or combination of symbols serving to express something

exterior angle
the angle outside a polygon formed between a line extending from a side and the adjacent side. The sum of the exterior angles of a convex polygon is 360°.

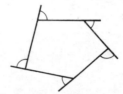

The exterior angles of a pentagon

extract
to find (a mathematical root) by calculation

extrapolate
to predict (a value of a variable at a given point) by extending a line or curve plotted on a graph from known values at previous points – compare INTERPOLATE

extremum (*plural* **extrema**)
a local maximum or minimum value of a mathematical function – see LOCAL MAXIMUM, TURNING POINT

face

any of the two-dimensional surfaces of a solid (eg a cube has six faces)

factor

1 any of the numbers or mathematical expressions that are multiplied together to form a product
2 an integer or mathematical expression that exactly divides another (eg 5 is a factor of 10; $(x-y)$ is a factor of x^2-y^2)
3 a quantity by which a given measurement is multiplied or divided in order to produce a measurement in terms of a different system of units
4 – see SCALE FACTOR

factor formulae

formulae used in trigonometry to express the sum or difference of a pair of sines or a pair of cosines as a product:

$$\sin A + \sin B = 2 \sin\tfrac{1}{2}(A + B)\cos\tfrac{1}{2}(A - B)$$
$$\sin A - \sin B = 2 \cos\tfrac{1}{2}(A + B)\sin\tfrac{1}{2}(A - B)$$
$$\cos A + \cos B = 2 \cos\tfrac{1}{2}(A + B)\cos\tfrac{1}{2}(A - B)$$
$$\cos A - \cos B = - 2\sin\tfrac{1}{2}(A + B)\sin\tfrac{1}{2}(A - B)$$

– compare ADDITION FORMULAE

factorial

a mathematical function of a positive integer, n, that is denoted by $n!$ and is equal to the result of multiplying together all the integers from 1 to n. Where the integer is zero, factorial zero, 0!, is given the value 1. $\langle n \sim \rangle$ $\langle \sim 3 \text{ is } 1 \times 2 \times 3 \rangle$ – see BINOMIAL THEOREM, COMBINATION

factorize *or* **factorise**

to express (a number or mathematical expression) in terms of its component factors

factor theorem

a theorem stating that if P(x) is a polynomial and P(a) = 0, then $(x - a)$ is a factor of P(x).

    Example
        P(x) = $x^3 + 5x^2 - 2x - 4$
        P(1) = $1 + 5 - 2 - 4 = 0$
    Hence $(x - 1)$ is a factor of P(x).
    [In fact $x^3 + 5x^2 - 2x - 4 = (x - 1)(x^2 + 6x + 4)$]

family
a set of curves, surfaces, etc that differ only by the constants appearing in their equations

F distribution
a probability density function that is used esp to compare the variances of two random variables that are independent and that both have a chi-square distribution

Fermat, Pierre (1601–65) French amateur mathematician who studied the theory of numbers

Fermat's last theorem
a theorem stating that the equation $a^n + b^n = c^n$ has no solution for integral values of a, b, and c, if $n > 2$

Fibonacci, Leonardo (c 12th century) of Pisa; nowadays remembered for the sequence of integers called Fibonacci's sequence

Fibonacci sequence
an infinite sequence of integers (eg 0, 1, 1, 2, 3, 5, 8, 13, ...) in which every number after the first two is the sum of the two numbers immediately preceding it. The Fibonacci sequence can be found in the form and development of many natural phenomena and is closely linked with the golden section in that the ratio of successive terms in the Fibonacci sequence tends to the golden section, $\frac{1}{2}(\sqrt{5} - 1) = 0·6180$, to 4 significant figures – see also GOLDEN SECTION

field
a set of mathematical elements (eg all the rational numbers) that when subject to the two binary operations of addition and subtraction is commutative under addition and, excluding zero (0), is commutative under multiplication. Within this set every nonzero element has an inverse and the distributive law applies.

figure
1 a number symbol; *esp* an arabic number symbol; a numeral, digit
2 a geometrical diagram or shape
– see also SIGNIFICANT FIGURES

finite
completely determinable in theory or in fact by counting or measurement

fit
the conformity between an experimental result and theoretical expectation or between data and an approximating curve – used esp in the phrase *goodness of fit*

fixed-point
involving or being a mathematical notation (eg in a decimal system) in which the point separating whole numbers and fractions is fixed – compare FLOATING-POINT

fixed point
a point which is left unchanged by a given mathematical transformation

floating-point
involving or being a mathematical notation in which a quantity is denoted by one number multiplied by a power of ten; (eg $230 \times 10^{-4} = 0.023 = 2.3 \times 10^{-2}$)
– compare FIXED-POINT, SCIENTIFIC NOTATION

fluxion (*archaic*)
CALCULUS 2

focal
of, having, or located at a focus

focus (*plural* **foci** *also* **focuses**)
a fixed point that together with a straight line (the directrix) forms a reference system for generating a curve (eg a parabola) that is a conic section – see CONIC SECTION, DIRECTRIX

foot (*abbr* **ft**)
an imperial unit of length originally based on the length of the human foot and equal to ⅓yd (0.3048m)

formula (*plural* **formulas** *or* **formulae**)
a general fact, rule, or principle expressed in symbols (eg the formula for the area of a circle is $A = \pi r^2$)

Fourier, Joseph (1768–1830) French mathematician who discovered a method for expressing periodic functions as an infinite series of trigonometrical functions

Fourier series
an infinite series in which the terms are constants multiplied by sine or cosine functions of integer multiples of the variable and which is used in the analysis of periodic functions (functions whose values repeat at intervals)

Example The periodic function $f(x) = x \ (-\pi \leqslant x < +\pi)$
$$f(x + 2\pi) = f(x)$$

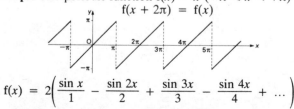

$$f(x) = 2\left(\frac{\sin x}{1} - \frac{\sin 2x}{2} + \frac{\sin 3x}{3} - \frac{\sin 4x}{4} + \ldots \right)$$

Fourier transform

Fourier transform
a function that under suitable conditions can be obtained from a given function of x by multiplying the given function by e^{-iux} and integrating over all values of x

fraction
a number (eg $\frac{3}{4}$, $\frac{5}{8}$, $0 \cdot 234$) that is the result of dividing two numbers. A proper fraction is a fraction less than 1 (eg $\frac{5}{8}$); an improper fraction is a fraction greater than 1 (eg $\frac{7}{3}$) – see PROPER FRACTION, IMPROPER FRACTION

French curve
a curved piece of flat material (eg plastic) used as an aid in drawing noncircular curves

frequency
1 the number of times that a periodic function repeats the same sequence of values as the independent variable varies by a specified amount
2 the number of things in a given section of an esp statistical distribution

frequency distribution
an arrangement of statistical data that shows the frequency of the occurrence of the values of a variable

frustum (*plural* **frustums** *or* **frusta**)
the part of a cone or pyramid formed by cutting off the top at a plane parallel to the base; *also* the part of a solid cut off by two usu parallel planes

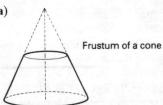

Frustum of a cone

F test
a statistical test using the F distribution

function
a mathematical rule that assigns to each element of a set (the domain) exactly one element of another set (the co-domain)
NOTATION $f(x) = x^2$, $f(5) = 25$
 or $f{:}x \mapsto x^2$, $f{:}5 \mapsto 25$

fundamental theorem of calculus
the theorem which shows that processes of integration (in particular finding the area under a curve) and differentiation (finding the gradient of a curve) are related

$$\lim_{\delta x \to 0} \sum_{x=a}^{x=b} f(x)\,\delta x = \int_a^b f(x)\,dx$$
$$= F(b) - F(a), \text{ where } f'(x) = f(x).$$

G

Galileo, Galilei (1564–1642) Italian astronomer and mathematician. Supporter of the Copernican Theory that the planets move around the sun, but forced to recant by the Inquisition.

gallon
an imperial unit of liquid capacity equal to 8 pints

Galois, Evariste (1811–32) French mathematician, one of the founders of modern algebra; tragically shot in a duel at the age of twenty-one

game theory *or* **games theory**
THEORY OF GAMES

Gauss, Carl (1777–1855) German, one of the greatest mathematicians of all time. He proved the fundamental theorem of algebra, that 'every polynomial equation, with complex coefficients, has at least one root in complex numbers'.

Gaussian distribution
NORMAL DISTRIBUTION

Gaussian integer
a complex number of the form $a + ib$, where a and b are real numbers

generator
1 a curve which, when rotated through 360°, traces the surface of a solid of revolution

Examples

(a) $y = \dfrac{r}{h} x$, generates a cone

(b) $y = \sqrt{(r^2 - x^2)}$, generates a sphere

2 an element of a group such that every element of the group can be expressed as one of its powers

geodesic *or* **geodesic line**
the shortest line between two points on a given surface

geometric mean
the nth root of the product of n numbers (eg the square root of two numbers) ⟨*the ~ of 9 and 4 is 6*⟩

geometric progression

geometric progression (*abbr* **GP**)
a sequence (eg 3, 6, 12, 24) in which the ratio of any term to its predecessor is constant

geometric series
a series ($a + ar + ar^2 + ar^3 + ...$) whose terms form a geometric progression (GP)
Formulae: The sum of the GP $a + ar + ar^2 + ar^3 + ... + ar^{n-1}$ is

$$\frac{a(r^n - 1)}{r - 1}$$

The sum of the (infinite) GP $a + ar + ar^2 + ar^3 + ...$ is

$$\frac{a}{1 - r} \text{ , provided } -1 < r < +1.$$

geometry
1 a branch of mathematics that deals with the measurement, properties, and relationships of points, lines, angles, surfaces, and solids; *broadly* the study of properties of given elements that remain constant under specified transformations
2 a particular type or system of geometry ⟨*Euclidean* ~⟩

Goldbach's conjecture
the proposition that every even number can be expressed as the sum of two primes (eg $12 = 5 + 7$)

golden section
the proportion of a geometric figure or of a divided line such that the smaller dimension is to the greater as the greater is to the whole. The golden section is found in many natural phenomena and is closely related to other mathematical forms (eg the logarithmic spiral) – see also FIBONACCI SEQUENCE

A   B   C
The golden section = AB/BC = BC/CA
$= \frac{1}{2}(\sqrt{5}-1) \approx 0.618$

goodness of fit
– see FIT

gradient
degree of slope, ie $\dfrac{\text{vertical displacement}}{\text{horizontal displacement}}$.

In Cartesian coordinates, the gradient of the line joining (x_1, y_1) and (x_2, y_2) is $\dfrac{y_2 - y_1}{x_2 - x_1}$

Example

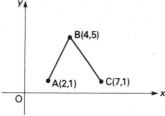

gradient of AB = $\dfrac{5-1}{4-2}$ = 2

gradient of BC = $\dfrac{1-5}{7-4}$ = $-\dfrac{4}{3}$

The gradient of the line $y = mx + c$ is m (eg the gradient of $y = 5x + 3$ is 5).

gradient function
DERIVED FUNCTION

gram *or* **gramme**
a metric unit of mass and weight equal to one thousandth of a kilogram (about 0·04oz)

graph
1 a diagram (eg a series of points, a line, a curve, or an area) expressing a relation between quantities or variables, usu having two (or three) axes with reference to which points or lines are located on the diagram
2 the collection of all points whose coordinates (numbers which specify the location of a point) satisfy a given relation (eg the equation of a mathematical function)

graphics
the art or science of drawing an object on a two-dimensional surface (eg the monitor of a microcomputer) according to mathematical rules of projection

graph paper
paper ruled for drawing graphs

great circle
a circle on the surface of a sphere whose centre is the centre of the sphere – compare LONGITUDE

Green, George (1793–1841) son of a Nottingham miller, self-taught mathematician, highly regarded by present day mathematical physicists

Gregory's series

Gregory's series
a series which expresses arc tan x in terms of powers of x. It may be used to evaluate π.

$$\arctan x = x - \frac{x^3}{3} + \frac{x^5}{5} - \frac{x^7}{7} + \ldots$$

special case ($x = 1$) gives

$$\frac{\pi}{4} = 1 - \frac{1}{3} + \frac{1}{5} - \frac{1}{7} + \ldots$$

which was the earliest method used for evaluating π.

gross (*plural* **gross**)
a group of twelve dozen things

group
a mathematical set which has an operation defined on pairs of elements of the set such that the operation is associative and which also has an identity element and an inverse for every element

Examples
(a) The set of even numbers under addition.
(b) The set of rotations, through 120°, of an equilateral triangle.
(c) The set $\{1, i, -1, -i\}$ under multiplication.

H

Halley, Edmund (1656–1742) English astronomer and mathematician; contemporary of Newton; calculated the orbit and predicted the return of the comet that now bears his name

hand
a unit of measure equal to 4in (about 102mm) used esp for the height of a horse

harmonic analysis
the expression of a periodic function (function whose value repeats itself at regular intervals) as a sum of sines and cosines and specif by means of a Fourier series

harmonic mean
the reciprocal of the arithmetic mean of the reciprocals of a finite set of numbers (eg the harmonic mean of 2, 4, 6, and 8 is

$$\frac{4}{\frac{1}{2} + \frac{1}{4} + \frac{1}{6} + \frac{1}{8}} = \frac{96}{25})$$

harmonic progression
a sequence whose terms are the reciprocals of an arithmetic progression (eg the sequence 1, ⅓, ⅕, ⅐)

harmonic series
a series constructed by adding together terms in a harmonic progression; *esp* the harmonic series 1 + ½ + ⅓ + ¼ + … (NB this series does not converge)

hect- *or* **hecto-**
hundred (10^2)

hectare
a metric unit of area equal to 10 000m²

height
the extent of elevation above a level; ALTITUDE. The height of a triangle is the length of a line through a vertex of a triangle and perpendicular to the opposite side. (The area of the triangle is ½ × base × height.)

helix
a curve traced on a cylinder by the rotation of a point moving up the cylinder at a constant rate; *broadly* SPIRAL 2

hemi-
half 〈hemi*sphere*〉

hemisphere
half a sphere

hepta- *or* **hept-**
seven

heptagon
a seven-sided polygon

Hermite, Charles (1822–1901) French mathematician who proved
that e is trancendental

Hero (or **Heron**) of Alexandria (c 1st century BC) Greek
geometer

Hero's formula
a formula for the area of a triangle whose sides are of lengths a, b
and c

$$A = \sqrt{s(s - a)(s - b)(s - c)} \text{ , where } s = \tfrac{1}{2}(a + b + c).$$

hexa- *or* **hex-**
six

hexadecimal
of or being a number system with a base of 16

> **Example** The hexadecimal number 235_{16} is equal to the decimal
> number $2 \times 16^2 + 3 \times 16 + 5 = 565$

hexagon
a six-sided polygon

hexagonal
1 having six angles and six sides
2 having a hexagon as section or base $\langle \sim prism \rangle$

highest common factor (*abbr* **HCF**)
the largest integer or the polynomial (series of algebraic terms) of
highest degree that can be divided exactly into each of two or
more integers or polynomials (eg the HCF of 42 and 54 is 6)

histogram
a graph consisting of rectangles whose bases represent equal
divisions of some continuous variable and whose heights represent
the number of occurrences of the variable within the division

homogeneity
in statistics the state of having identical distribution functions or
values $\langle a \text{ test for } \sim \text{ of variances} \rangle$ $\langle \sim \text{ of two statistical populations} \rangle$

homogeneous
of an equation, fraction, etc having each term of the same degree if
all variables are considered, eg $x^2 + xy + y^2 = 0$ is a
homogeneous equation of degree 2. A homogeneous differential

equation (in x and y) is a differential equation in which the terms in x and y are homogeneous, eg

$$x^2\frac{dy}{dx} = 3xy - 5y^2$$

homomorphism
a mathematical function that maps elements of one mathematical set onto elements of another and that has the property that applying the function to the sum or product of two elements is equivalent to applying the function to each element separately and adding or multiplying the result

Example The mapping $x \mapsto \log x$
$$A \mapsto \log A$$
$$B \mapsto \log B$$
and $AB \mapsto \log (AB) = \log A + \log B$

horizontal
parallel to, or in the plane of, a plane parallel to the horizon or to a base line

horsepower
an imperial unit of power equal to 745·7 watts

hundredweight
a unit of weight equal to 112lb (about 50·80kg)

hyper-
that exists in, or is a space of, more than three dimensions ⟨hyper*cube*⟩ ⟨hyper*space*⟩

hyperbola
the locus of a point P which moves so that its distance from a fixed point S (the focus) is in a constant ratio e (the eccentricity) to its distance from a fixed line l (the directrix),

$$\frac{PS}{PM} = e, \text{ where } e > 1.$$

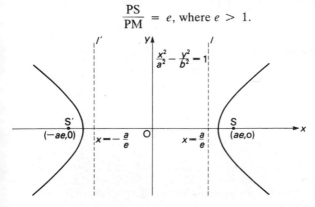

hyperbolic

The equation of the hyperbola referred to axes through O is

$$\frac{x^2}{a^2} - \frac{y^2}{b^2} = 1, \text{ (where } b^2 = a^2(e^2 - 1))$$

If $a = b$, the curve is called a rectangular hyperbola. The equation of a rectangular hyperbola when its asymptotes are the axes is $xy = c^2$.

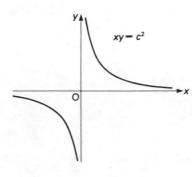

A hyperbola can alternatively be thought of as a two-dimensional curve generated by a point so moving that the difference of its distances from two foci is a constant; the intersection of two right circular cones joined at the vertices with a plane that cuts both of the cones – compare CONIC SECTION, ELLIPSE, PARABOLA

hyperbolic
1 of or like a hyperbola
2 of or being a space in which more than one line parallel to a given line passes through a point ⟨~ *geometry*⟩

hyperbolic function
any of a set of six mathematical functions that are related to the hyperbola and are analogous to the trigonometric functions: hyperbolic sine (sinh), hyperbolic cosine (cosh), hyperbolic tangent (tanh), and their reciprocals hyperbolic cosecant (cosech), hyperbolic secant (sech), and hyperbolic cotangent (coth)

$$\begin{aligned}
\text{Definitions: } \sinh x &= \tfrac{1}{2}(e^x - e^{-x}) \\
\cosh x &= \tfrac{1}{2}(e^x + e^{-x}) \\
\tanh x &= \frac{\sinh x}{\cosh x} = \frac{e^x - e^{-x}}{e^x + e^{-x}}
\end{aligned}$$

hyperboloid
a curved surface generated by an ellipse of variable size moving in such a way that it touches the curves of two equal hyperbolas whose planes are perpendicular to each other and which have one axis in common. The hyperboloid may have either of two different forms, one of a single somewhat hourglass-shaped surface, the

other of two surfaces stretching to infinity; the cross sections of the hyperboloids are either hyperbolas or ellipses. – compare
ELLIPSOID, PARABOLOID

hyperspace
space of more than three dimensions

hypocycloid
a curve traced by a point on the circumference of a circle that rolls round the inside of a fixed circle

hypotenuse
the side of a right-angled triangle that is opposite the right angle

hypothesis
a (plausible) proposition used to explain or generalize certain facts. Once it has been proved rigorously, it is called a theorem.

I

i
1 (*in roman type*: i) the square root of −1; $\sqrt{-1}$
2 (*in bold type*: **i**) a unit vector parallel to the *x*-axis

I
the symbol for identity matrix

icosahedron (*plural* **icosahedrons** *or* **icosahedra**)
a solid having 20 faces. A regular icosahedron has 20 equal
equilateral triangular faces.

identity
1 an algebraic equation (eg $(x + y)^2 = x^2 + 2xy + y^2$) that
remains true whatever values are substituted for the symbols
2 IDENTITY ELEMENT

identity element
an element that leaves any element of the set to which it belongs
unchanged when combined with it by a specific mathematical
operation

Examples
(a) 0 is the identity element under addition of real numbers.
(b) 1 is the identity element under multiplication of real
numbers.

(c) $\begin{pmatrix} 1 & 0 \\ 0 & 1 \end{pmatrix}$ is the identity element under multiplication of

2×2 matrices.

identity matrix (*symbol* I)
a square matrix such that $\mathbf{MI} = \mathbf{IM} = \mathbf{M}$; an identity matrix has 1s
on the principal diagonal (top left to bottom right) and 0s
elsewhere,

eg $\begin{pmatrix} 1 & 0 & 0 \\ 0 & 1 & 0 \\ 0 & 0 & 1 \end{pmatrix}$ is the 3 × 3 identity matrix – called also
UNIT MATRIX

image
an element in a set (the range) onto which an element of another
set (the domain) is mapped by a mathematical function

Example For the function f:$x \mapsto x^2$, the image of 5 is 25

image set
RANGE 2

imaginary number

a complex number of the form bi, where b is a real number

imaginary part

the part of a complex number that contains the imaginary number. The imaginary part of a complex number $z = a + ib$ (written $\text{Im}(z)$) is b (eg if $z = 4 + 3i$, then $\text{Im}(z) = 3$) – compare REAL PART

NB Two complex numbers $a + ib$ and $c + id$ are equal if and only if their real parts are equal and their imaginary parts are equal, ie $a = c$ and $b = d$

imperial unit

a unit belonging to an official nonmetric British series of weights and measures

imperial gallon

a unit of liquid capacity used as a standard in Britain and equal to about 4·546 litres

implicit function

a mathematical function that cannot be expressed with the dependent variable on one side of an equation and the one or more independent variables on the other (eg $x^3 + y^3 = \sin(xy)$ cannot be arranged as an expression for y in terms of x) – compare EXPLICIT FUNCTION

improper fraction

1 *in arithmetic* a fraction whose numerator is equal to or larger than the denominator (eg $\frac{7}{3}$)

2 *in algebra* a fraction whose numerator is of equal or higher degree than the denominator

$$(\text{eg } \frac{x^3 + 1}{x + 2})$$

– compare PROPER FRACTION

improper integral

a definite integral (a) whose region of integration includes a point at which the integrand (quantity to be integrated) is undefined or tends to infinity

$$(\text{eg } \int_0^1 \frac{1}{\sqrt{(1 - x^2)}} \, dx)$$

or (b) whose limits of integration are not both finite

$$(\text{eg } \int_1^\infty 1/x^2 \, dx)$$

incentre
– see INCIRCLE

inch
a unit of length equal to ⅟₃₆yd (about 25·4mm)

incircle
the circle which touches the
three sides of a given triangle.
The centre of the circle (the
incentre) is found by bisecting
the angles of the triangle.

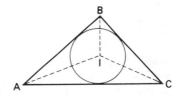

inclination
1 the angle between two lines or planes
2 the angle made by a line with the *x*-axis measured anticlockwise
from the positive direction of that axis

inclined plane
a plane surface that makes an oblique angle with the horizontal

included angle
an angle bounded by two given
sides of a triangle

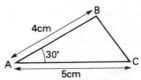

Angle A is the included angle
(between sides AB and AC)

inconsistent
not satisfiable by any set of values for the unknowns ⟨~
equations⟩ ⟨~ *inequalities*⟩ – compare INDETERMINATE

    Examples
     (a) The equations $x + y = 1$ and $x + y = 5$.
     (b) The equations $x + y + z = 1$, $x - y + z = 3$ and
$x + z = 7$.

increment
a small positive or negative change in the value of a mathematical
quantity; *esp* a minute increase in a mathematical quantity – see
DELTA

indefinite integral
a function whose derivative is a given function – called also
PRIMITIVE

independent

Examples

f(x)	$\int$f(x) dx (constants omitted)
x^n	$\dfrac{x^{n+1}}{n+1}$, $(n \neq -1)$
x^{-1}	$\ln x$
$\sin x$	$-\cos x$
$\cos x$	$\sin x$
$\sec^2 x$	$\tan x$
e^x	e^x

independent

1 having linear independence ⟨*an ~ set of vectors*⟩
2 having the property that the joint probability (eg of events or samples) equals the product of the separate probabilities
3 neither deducible from, nor incompatible with, another statement ⟨*~ postulates*⟩

independent variable

any of an arbitrary set of variables in an equation whose values may be freely chosen (eg in $z = x^2 + 3xy + y^2$, x and y may be taken as independent variables) – compare DEPENDENT VARIABLE, DUMMY VARIABLE

indeterminate

of simultaneous equations having an infinite number of solutions by reason of certain relations between the coefficients – compare INCONSISTENT

Example The equations $x + y + z = 1$, $x - y + z = 3$ and $x + z = 2$

index (*plural* indices)

POWER 1
Rules of indices:

$$x^a \times x^b = x^{a+b}$$
$$x^a \div x^b = x^{a-b}$$
$$(x^a)^b = x^{ab}$$

Special cases:

$$x^{-a} = 1/x^a,\ x^0 = 1,\ x^{1/a} = {}^a\sqrt{x}$$

indirect proof

REDUCTIO AD ABSURDUM

indivisible

not divisible

induction

a mathematical demonstration of the validity of a law concerning positive integers, by proving that the law holds for the first integer and that if it holds for an integer k, then it must also hold for $k + 1$

inequality
a formal statement of mathematical inequality between two expressions (eg $x < 5$ (x is less than 5))

inference
the assumption of statistical generalizations (eg of the value of population parameters) usu with calculated degrees of certainty, based on actual sample values determined by experiment

infinite
1 greater than any arbitrarily chosen finite value, however large ⟨*there is an ~ number of positive integers*⟩
2 extending to infinity ⟨*~ plane surface*⟩
3 having an infinite number of elements or terms ⟨*an ~ set*⟩ ⟨*an ~ series*⟩

infinitesimal
taking on values arbitrarily close to zero

infinitesimal calculus
CALCULUS 2

infinity
1 an arbitrarily large number. Symbol ∞
2 a part of a geometric figure that lies beyond any part whose distance from a given position is finite ⟨*do parallel lines ever meet if they extend to ~?*⟩

inflection *also* (*in Britain*) **inflexion**
change of curvature with respect to a fixed line from concave to convex or conversely

A point of inflection on $y = f(x)$ is a point where d^2y/dx^2 is zero and changes sign.

> **Example** $y = 2x^3 - 6x^2 + 6$
>
> $$\frac{dy}{dx} = 6x^2 - 12x$$
>
> $$\frac{d^2y}{dx^2} = 12x - 12 = 12(x - 1)$$

The second derivative is zero and changes sign when $x = 1$, so there is a point of inflection at (1,2).
In contrast, consider $y = x^4$ at (0,0).

inner product

SCALAR PRODUCT

inscribe

to draw within a geometrical figure so as to touch at as many points as possible ⟨*a regular polygon* ~d *in a circle*⟩

integer

the number one (1) or any number obtainable by once or repeatedly adding one to or subtracting one from the number one; a whole number (eg ... -3, -2, -1, 0, 1, 2, 3, ...)

NOTATION the set of integers $= \mathbb{Z}$

– compare REAL NUMBER, RATIONAL NUMBER

integral (*adjective*)

1 of or being a mathematical integer

2 of or concerned with mathematical integrals or integration

integral (*noun*)

1 a mathematical function whose derivative is a given function – compare DEFINITE INTEGRAL, INDEFINITE INTEGRAL

2 a solution of a differential equation

integral calculus

a branch of mathematics dealing with methods of finding integrals and with their applications (eg to the determination of lengths, areas, and volumes and to the solution of differential equations)

integrand

a mathematical expression to be integrated

integrate

to find the integral of (eg a mathematical function or differential equation)

integration

the reverse of differentiation; the operation of calculating an integral

integration by parts

a method of mathematical integration using the formula

$$\int u\left(\frac{\mathrm{d}v}{\mathrm{d}x}\right) \mathrm{d}x = uv - \int v\left(\frac{\mathrm{d}u}{\mathrm{d}x}\right) \mathrm{d}x$$

intercept

the distance from the zero point (origin) on a graph to another point where the graph crosses one of the axes. The intercept form of the equation of the straight line through $(a,0)$ and $(0,b)$ is

$$\frac{x}{a} + \frac{y}{b} = 1.$$

interest

the money paid or payable when a sum of money is invested or borrowed. If

P = sum invested (the principal)
R = the (annual) rate of interest
N = period of time in years
S = value of investment after N years simple interest
    (ie the interest not reinvested)
C = value of investment after N years compound interest
    (ie with interest reinvested)

$$S = P + \frac{P \times R \times N}{100} \qquad C = P\left(1 + \frac{R}{100}\right)^N$$

Example £1000 invested at 8% for 10 years

$$S = 1000 + \frac{1000 \times 8 \times 10}{100} = £1800$$

$$C = 1000 \times (1 \cdot 08)^{10} = £2160$$

interior angle

an angle contained within any two adjacent sides of a polygon

interpolate

to estimate values that lie between two known values of (a curve or other mathematical function) – compare EXTRAPOLATE

interquartile range

the difference between the upper quartile and the lower quartile in a frequency distribution – see QUARTILE

intersect

to pierce or divide (eg a line or area) by passing through or across; cross

intersection

1 the set of points where two lines meet (and cross)

2 the set of elements common to two or more mathematical sets – see CAP; compare UNION

NOTATION the intersection of sets A and B = A ∩ B

Example
  A = {a, b, c, d}
  B = {c, d, e}
  A ∩ B = {c, d}
The intersection of A and B is represented by the shaded region in the Venn diagram.

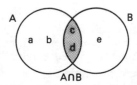

interval

a set of real numbers between two numbers, that either includes or excludes one or both of them

NOTATION The set of values of x such that $a \leqslant x \leqslant b$ is the *closed* interval $[a, b]$.

The set of values of x such that $a < x < b$ is the *open* interval (a, b).

invariant

constant, unchanging; *specif* unaffected by a particular mathematical operation

inverse

1 *of a fraction* the reciprocal (eg the inverse of a/b is b/a)

2 *of a function* a function which reverses the effect of a given function (eg the inverse of $f(x) = 2x + 1$ is $f^{-1}(x) = \frac{1}{2}(x - 1)$; $f(5) = 11$, $f^{-1}(11) = 5$)

3 *of a matrix M* a matrix $\mathbf{M}^{-1}$, such that $\mathbf{MM}^{-1} = \mathbf{M}^{-1}\mathbf{M} = \mathbf{I}$, the identity matrix. The inverse of the 2×2 matrix

$$\begin{pmatrix} a & b \\ c & d \end{pmatrix} \text{ is } \frac{1}{ad - bc} \begin{pmatrix} d & -b \\ -c & a \end{pmatrix}$$

provided $ad - bc \neq 0$.

inverse proportion

the relationship between two variables in which one increases in the same proportion as the other decreases so that their product is constant; y is inversely proportional to x^n, means that y is proportional to $1/x^n$, ie $y = k/x^n$ – compare DIRECT PROPORTION, VARIATION

involute

the curve that would be traced by a point on a thread kept taut as it is unwound from another curve (the evolute)

irrational number

a number (eg $\sqrt{2}$, π, e) that cannot be expressed as the ratio of two integers – compare RATIONAL NUMBER, SURD

irreducible

of a mathematical expression incapable of being split up into factors whose variables have lower powers (than those in the expression) (eg $x^2 + 1$)

isometric

relating to or being an isometric drawing or isometric projection

isometric drawing

a three-dimensional representation of an object in which the lines in all three dimensions are drawn to scale

isometric projection

a method of drawing a three-dimensional object in which the three faces shown are equally inclined to the drawing surface so that all the edges are equally foreshortened

isometry

a mapping of a geometrical figure so that the distance between any two points in the original figure is the same as the distance between their images in the second. Rotation and translation are (direct) isometries; reflection is an opposite isometry.

isomorphism

a relationship between the elements of two groups which shows that the groups have the same structure

Example

	a	b	c	d
a	a	b	c	d
b	b	a	d	c
c	c	d	b	a
d	d	c	a	b

	e	x	y	z
e	e	x	y	z
x	x	e	z	y
y	y	z	x	e
z	z	y	e	x

The isomorphism is $a \leftrightarrow e$, $b \leftrightarrow x$, $c \leftrightarrow y$, $d \leftrightarrow z$.

isosceles triangle

a triangle having two sides of equal length

iteration

an iterative procedure

iterative

relating to or being an arithmetical procedure in which repetition of a cycle of operations produces results which are ever closer approximations to an unknown value – see also ALGORITHM

Example $x_{n+1} = \frac{1}{2}(x_n + 17/x_n)$, $x_1 = 4$

is an iterative formula for evaluating $\sqrt{17}$

$x_1 = 4$
$x_2 = \frac{1}{2}(4 + 17/4) = 4 \cdot 125$
$x_3 = \frac{1}{2}(4 \cdot 125 + 17/4 \cdot 125) \approx 4 \cdot 123$
$x_4 = \frac{1}{2}(4 \cdot 123 + 17/4 \cdot 123) \approx 4 \cdot 123$
Hence $\sqrt{17} \approx 4 \cdot 123$.

J

j
1 (*in roman type*: j) – sometimes used, esp by electrical engineers, for $\sqrt{-1}$
2 (*in bold type*: **j**) a unit vector parallel to the y-axis

Jordan curve
SIMPLE CLOSED CURVE (eg a circle or ellipse)

joule
the SI unit of work or energy equal to the work done when a force of one newton moves its point of application through a distance of one metre

K

k
a unit vector parallel to the z-axis

kelvin (*symbol* **K**)
the SI unit of temperature defined by the Kelvin scale on which absolute zero (the hypothetically coldest temperature possible) is at 0K and water freezes at 273·16K

Kepler, Johann (1571–1630) astronomer whose work paved the way for Newton's universal law of gravitation

kilo-
thousand (10^3)

kilogram (*abbr* **kg**)
the SI unit of mass

kilogram-weight (*abbr* **kg-wt**)
a unit of force equal to the weight of a kilogram mass under the earth's gravitational attraction – compare NEWTON

kilometre (*abbr* **km**)
a unit of length equal to 1000m (about 0·62mi)

kilowatt (*abbr* **kW**)
a unit of power equal to 1000 watts

kinematics
the study of motion (without reference to force and mass); *esp* the study of the relationship between displacement, velocity, and acceleration

kinetic energy
energy due to motion. The kinetic energy of a particle of mass m moving with velocity v is $\frac{1}{2}mv^2$ – compare POTENTIAL ENERGY, WORK DONE

kite
a quadrilateral with two pairs of adjacent equal sides – compare RHOMBUS

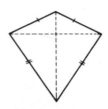

Klein bottle

Klein bottle

a one-sided surface that is formed by passing the narrow end of a tapered tube through the side of the tube and flaring this end out to join the other end – compare MÖBIUS STRIP

Klein group

the only noncyclic group of order 4

	e	x	y	z
e	e	x	y	z
x	x	e	z	y
y	y	z	x	e
z	z	y	e	x

knot

an imperial unit of speed equal to 1 nautical mile per hour – see NAUTICAL MILE

Kronecker delta

a function of two variables that is defined to be equal to one (1) when the variables are equal and zero (0) otherwise

kurtosis

the peakedness or flatness of the graph of a frequency distribution, esp as determining the concentration of values near the mean as compared with the normal distribution

L

Lagrange Joseph-Louis (1736–1813) French mathematician (but born in Turin), author of *Mécanique Céleste*

Lagrange's theorem
a theorem stating that the order of a subgroup of a finite group is a divisor of the order of the group

Latin square
a square array of symbols in which each symbol appears once and once only in each row and column and which is used in the design of statistical experiments. If the symbols also obey the associative law, then the Latin square represents a group.

Example

	A	B	C
A	A	B	C
B	B	C	A
C	C	A	B

latitude
angular distance of a point on the earth's surface measured north or south from the equator. A circle of latitude is a circle on the earth's surface parallel to the equator – compare LONGITUDE

lattice
a regular geometrical arrangement of points over an area or in space

latus rectum
a straight line passing through a focus of a conic section parallel to the directrix

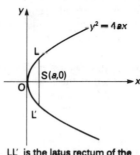

LL′ is the latus rectum of the parabola $y^2 = 4ax$

law
a relation between mathematical or logical expressions that holds for all cases

leading diagonal
PRINCIPAL DIAGONAL

least squares
a method of fitting a curve to a set of points representing statistical data, in such a way that the sum of the squares of the distances of the points from the curve is a minimum

Leibnitz, Gottfried (1646–1716) German mathematician, codiscoverer (with Newton) of calculus

Leibnitz' notation
the notation dy/dx, d^2y/dx^2, etc, which is attributed to Leibnitz

lemma
a minor theorem, used as part of a complex proof

lemniscate
a figure-of-eight shaped curve whose equation in polar coordinates is $r^2 = a^2 \cos 2\theta$

level of significance
the probability that a given statistical outcome could have arisen by chance rather than as a result of some specific cause

light-year
a unit of length in astronomy equal to the distance that light travels in one year in a vacuum; 9460 thousand million kilometres (about 5878 thousand million miles)

limaçon
a geometric curve which can occur in three forms, one of which is somewhat heart-shaped, and that has the equation in polar coordinates $r = b + 2a \cos \theta$ – see CARDIOID

limit
1 a number whose difference from the value of a function approaches zero as the value of the independent variable approaches some given number (eg the limit of $\sin\theta/\theta$ as $\theta \to 0$, is 1)

2 a number that for an infinite sequence of numbers is such that ultimately each of the remaining terms of the sequence differs from this number by less than any given amount (eg the limit of the sequence 1, $\frac{1}{2}$, $\frac{1}{4}$, $\frac{1}{8}$, $\frac{1}{16}$, ... is 0)

line
the locus of a point which traces the shortest distance between two given points and which continues indefinitely in either direction – see SEGMENT

linear
1 of or resembling a line; straight
2 involving a single dimension

3 of or being a relationship between a pair of variables which can be expressed by a linear equation (eg $y = mx + c$)

linear algebra
a branch of mathematics concerned with vector spaces and matrices

linear combination
a mathematical expression (eg $4x + 5y + 6z$) which is composed only of additions or subtractions of elements (eg variables or vectors)

linear equation
an equation of the first degree in any number of variables (eg $x + y = 9$)

linear programming
a mathematical method of solving practical problems (eg the most profitable allocation of resources) by finding the maximum or minimum value of linear functions subject to various constraints

linearly dependent
a set of vectors **a**, **b**, **c**, ... is linearly dependent if there exists a set of real numbers α, β, γ, ... (not all zero) such that
$$\alpha\mathbf{a} + \beta\mathbf{b} + \gamma\mathbf{c} + ... = \mathbf{0}$$
 Example a = $2\mathbf{i} + 2\mathbf{j}$, **b** = $3\mathbf{i} + 3\mathbf{j} + 2\mathbf{k}$, **c** = **k**,
$3\mathbf{a} - 2\mathbf{b} + 4\mathbf{k} = \mathbf{0}$

linearly independent
not linearly dependent

litre
a metric unit of capacity equal to 1 cubic decimetre ($1dm^3$)

local maximum (*or* **minimum**)
a function $f(x)$ is said to have a local maximum (or minimum) when $x = a$, if $f(a)$ is greater (or less) than $f(x)$ for values of x near a.

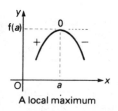
A local maximum

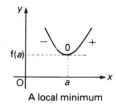
A local minimum

At a local maximum the gradient $f'(x)$ changes from positive, through zero, to negative; at a local minimum the gradient changes from negative, through zero, to positive – see also
TURNING POINT, STATIONARY POINT, POINT OF INFLECTION

local minimum

local minimum
– see LOCAL MAXIMUM

locus (*plural* **loci** *also* **locuses**)
the set of all points whose location is determined by stated
conditions (eg the locus of a point which moves so that its distance
from a fixed point O is constant, is a circle centre O)

logarithm
the power to which a number (the base) must be raised in order to
produce a given number ⟨*the ~ of 100 to base 10 is 2*⟩
Rules of logarithms:

$$\log a + \log b = \log(ab)$$
$$\log a - \log b = \log(a/b)$$
$$m\log a = \log(a^m)$$

Special cases:

$$\log 1 = 0$$
$$\log(1/a) = -\log a$$
$$\tfrac{1}{2}\log a = \log\sqrt{a}$$

– see also COMMON LOGARITHM, NATURAL LOGARITHM

logarithmic function
a function (eg $y = \log_a x$) that is the inverse of an exponential
function (eg $x = a^y$)

longitude
angular distance of a point on the earth's surface measured east or
west along the equator from the Greenwich meridian. A circle of
longitude is a great circle passing through the poles – compare
LATITUDE, MERIDIAN

lowest common denominator *or* **least common denominator**
(*abbr* **LCD**)
the lowest common multiple of two or more denominators (eg the
LCD of the fractions ⅓, ⅕, ⅙ is 30)

lowest common multiple (*abbr* **LCM**)
the smallest integer which is a multiple of two or more other
integers (eg the LCM of 3,5,6 is 30)

lowest terms
the form of a fraction in which the numerator and denominator
have no factor in common (eg the fraction 6/9 reduced to its lowest
terms is ⅔)

Maclaurin, Colin (1698–1746) Scottish mathematician whose main interest was geometry, but today he is usually remembered for Maclaurin's series

Maclaurin's series
a Taylor's series in which the expansion is about the reference point zero

Examples
$$\sin x = x - \frac{x^3}{3!} + \frac{x^5}{5!} - \frac{x^7}{7!} + \ldots$$
$$\cos x = 1 - \frac{x^2}{2!} + \frac{x^4}{4!} - \frac{x^6}{6!} + \ldots$$
$$e^x = 1 + \frac{x}{1!} + \frac{x^2}{2!} + \frac{x^3}{3!} + \ldots$$
$$\ln(1 + x) = x - \frac{x^2}{2} + \frac{x^3}{3} - \frac{x^4}{4} + \ldots$$

magic square
a square array of numbers in which the numbers in each row, column, and diagonal add up to the same total

Example

2	9	4
7	5	3
6	1	8

major
larger; *esp* being the larger of two parts ⟨~ *arc of a circle*⟩ ⟨~ *sector of a circle*⟩ – compare MINOR

major axis
the longest axis of an ellipse; the chord passing through the focuses of an ellipse

mantissa
the part of a common logarithm (logarithm whose base is 10) following the decimal point (eg ·5359 of the logarithm 3·5359) – compare CHARACTERISTIC

mapping
a correspondence between two mathematical sets in which each element of one set corresponds exactly to one element of the other set

Markov chain *also* **Markoff chain**
a sequence of random events or states in which the probability of

mathematics

each event or state occurring is dependent on the outcome of the preceding event

mathematics
the science of numbers and their operations, interrelations, combinations, generalizations, and abstractions and of space configurations and their structure, measurement, transformations, and generalizations

matrix
a rectangular array of numbers, eg $\begin{pmatrix} 1 & 3 & 5 \\ 4 & 7 & 9 \end{pmatrix}$

usu enclosed in round brackets (but square brackets may also be used) – see also ELEMENTARY OPERATION

maximum
– see LOCAL MAXIMUM

mean
1 ARITHMETIC MEAN.
The arithmetic mean of a set of n numbers $x_1, x_2, x_3, \ldots x_n$ is
$$(x_1 + x_2 + x_3 + \ldots + x_n)/n.$$
The arithmetic mean of a frequency distribution in which the value x_i occurs with frequency f_i is
$$(f_1x_1 + f_2x_2 + f_3x_3 + \ldots + f_nx_n)/N$$
where $N = f_1 + f_2 + f_3 + \ldots + f_n$.
– see also GEOMETRIC MEAN, HARMONIC MEAN
2 The mean of a (continuous) function f(x) over an interval $x = a$ to $x = b$ is

$$\frac{1}{(b - a)} \int_a^b \mathrm{f}(x) \, \mathrm{d}x$$

mechanics
the study of certain phenomena (eg velocity, acceleration, force, energy, momentum) by applying mathematical techniques to physical laws, esp Newton's laws of motion – see also DYNAMICS, STATICS

median
1 a statistical value in an ordered set of values below and above which there is an equal number of values or which is the arithmetic mean of the two middle values if there is no one middle value (eg the median of 1,2,3,7,8,9,10 is 7)
2 a line from a vertex of a triangle to the midpoint of the

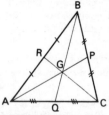

AP, BQ, CR are medians of triangle ABC

opposite side. The medians of a triangle meet at the centroid of the triangle.

mediator
a line of symmetry

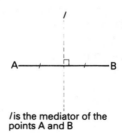

l is the mediator of the points A and B

mega- *or* meg-
million (10^6) ⟨mega*hertz*⟩ ⟨mega*watt*⟩

member
ELEMENT 1. $6 \in$ {even numbers} means '6 is a member of the set of even numbers'.

mensuration
1 the act of measuring; measurement
2 geometry applied to the estimation of lengths, areas, or volumes from given dimensions or angles

meridian
1 a great circle on the surface of the earth passing through the geographic poles; *also* the half of such a circle included between the poles
2 a representation of such a circle or half circle numbered for longitude on a map or globe (eg the Greenwich meridian is 0° longitude) – see also LONGITUDE
3 **meridian** *or* **meridian curve** the curve formed by the intersection of the surface of a revolving body and a plane passing through the axis of revolution

Mersenne prime
a prime number of the form $2^p - 1$, where p is a prime number (eg $31 = 2^5 - 1$).
NB not all numbers of this form are prime (eg $2^{67} - 1$).

metre (*abbr* m)
the SI unit of length equal to 1 650 763·73 wavelengths of the radiation corresponding to the transition between two specific energy levels of the krypton isotope $_{36}Kr^{86}$

metric system
a decimal system of weights and measures based on the metre and on the kilogram

metric ton
TONNE (1000kg)

micro-
one millionth (10^{-6}) part of (a specified unit) (eg *microgram*: one millionth of a gram; *microsecond*: one millionth of a second)

midpoint
a point midway between the ends of a line segment

midpoint theorem
a theorem stating that the line joining the midpoints of two sides of a triangle is parallel to the third side and equal to half of it in length

mile
an imperial unit equal to 1760yd (about 1·61km)

millenium
a period of 1000 years

milli-
one thousandth (10^{-3}) part of (a specified unit) (eg *milligram*: one thousandth of a gram; *millilitre*: one thousandth of a litre; *millimetre*: one thousandth of a metre; *millisecond*: one thousandth of a second)

minimum
– see LOCAL MAXIMUM

minor (*noun*)
a determinant obtained from a given determinant by eliminating the row and column in which the given element lies

Example In the determinant

$$\begin{vmatrix} 1 & 3 & 2 \\ 5 & 7 & 1 \\ 1 & 6 & 4 \end{vmatrix}$$

the minor of the element 2 is the determinant

$$\begin{vmatrix} 5 & 7 \\ 1 & 6 \end{vmatrix} = 5 \times 6 - 7 \times 1 = 23.$$

minor (*adjective*)
smaller; *esp* being the smaller of two parts $\langle \sim$ *arc of a circle* $\rangle$ $\langle \sim$ *sector of a circle* $\rangle$ – compare MAJOR

minor axis
the chord of an ellipse that passes through the centre and is perpendicular to the major axis

minute
the 60th part of an hour of time or of a degree of circular
measurement

mixed number
a number (eg 5⅔) composed of an integer and a proper fraction

Möbius strip
a surface having only one side
and one edge that can be
constructed from a rectangle by
holding one end fixed, rotating
the opposite end through 180°,
and joining it to the first end

mode
the most frequently occurring value in a set of data (eg the mode
of 1, 1, 2, 2, 2, 3, 4, 4, 5 is 2)

model
a system of postulates, data, inferences, or equations presented as
a mathematical description of a physical situation

modular arithmetic
arithmetic that deals with integers where the numbers are replaced
by their remainders after division by the modulus (eg 3 multiplied
by 4 in a modular arithmetic with modulus 5, would be 2)

modulo
with respect to a modulus of ⟨ *19 and 54 are congruent* ~ *7* ⟩.
Arithmetic modulo n uses the integers 0, 1, 2, 3 ... $(n - 1)$.

modulus
1 the magnitude of a real number irrespective of sign; the
modulus of x is written $|x|$. If $x \geq 0$, then $|x| = x$, if $x < 0$, then
$|x| = -x$ (eg $|+3| = 3$, $|-5| = 5$, $|-7 \cdot 5| = |+7 \cdot 5| = 7 \cdot 5$)
2 the positive square root of the sum of the squares of the real and
imaginary parts of a complex number, ie the modulus of a complex
number $z = a + ib$, written $|z|$ is $\sqrt{(a^2 + b^2)}$
(eg $|3 + 4i| = \sqrt{(3^2 + 4^2)} = 5$)
3 an integer that divides without remainder the difference
between two other integers

moment
(a measure of) the tendency to produce rotational movement
about a point or axis. The moment of a force F about a point O, is
$p \times F$, where p is the perpendicular distance from O to the line of
action of F – see TORQUE

monotonic
having either of the properties of always increasing or always
decreasing (eg f$(x) = x^3$ is a monotonic function)

Monte Carlo

of or involving the use of random sampling techniques to obtain approximate solutions to mathematical or physical problems, esp in terms of a range of values each of which has a calculated probability of being the solution ⟨~ *methods*⟩

multiple

the result of multiplying a quantity by a whole number (eg 35 is a multiple of 7)

N

nano-
one thousand millionth (10^{-9}) part of (a specified unit) (eg
nanometre: one thousand millionth of a metre; *nanosecond*: one
thousand millionth of a second)

Napier, John (1550–1617) compiler of the earliest set of (natural)
logarithm tables

Napierian logarithm
NATURAL LOGARITHM

natural logarithm (*symbol* **ln**)
a logarithm with e = approx. 1·71828 as base

natural numbers
the numbers 1, 2, 3, 4, ... (sometimes 0 is included)
NOTATION $\mathbb{N} = \{1, 2, 3, 4, ...\}$

nautical mile
any of various units of distance used for sea and air navigation
based on the length of a minute (unit of angular distance) of arc of
a great circle on the earth's surface and differing because the earth
is not a perfect sphere. The British nautical mile is equal to 6080ft
(about 1853·18m) and the international nautical mile (adopted by
the UK in 1970) is equal to 1852m (about 6076·17ft). 1 knot is
equal to 1 nautical mile per hour.

necessary condition
a proposition in logic or mathematics whose falsity assures the
falsity of another – compare SUFFICIENT CONDITION

> **Example** For an integer (other than 2) to be a prime number it
> is *necessary*, but not *sufficient*, that it should be odd.

negative
1 numerically less than zero; opposite in sign to a positive number
⟨−2 *is a* ~ *number*⟩

2 extending or generated in a direction opposite to the positive
direction

3 *of an angle* measured in a clockwise direction

neighbourhood
the set of all points whose distances from a given point are not
greater than a given (arbitrarily small) positive number

net of a solid
a plane diagram, showing all
the faces of a solid, which can
be folded to construct the solid

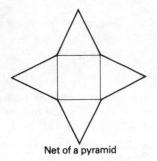

Net of a pyramid

newton
the SI unit of force equal to the force that will give an acceleration
of 1 metre per second per second to a mass of 1 kilogram

Newton, Sir Isaac (1642–1727) British mathematician and
physicist. Codiscoverer (with Leibnitz) of calculus (first published
by Newton in *Principia Mathematica* in 1687). He also formulated
'Newton's Laws of Motion' and made many important scientific
discoveries. Sometime Master of the Royal Mint. Regarded by
many as the greatest mathematician of all time.

Newton-Raphson method
an iterative method for solving an equation of the form f(x) = 0,
when an approximate root a_0 is known; further approximations are
given by

$$a_{n+1} = a_n - \frac{f(a_n)}{f'(a_n)}$$

nilpotent
equal to zero when raised to some power $\langle \sim matrices \rangle$

Example

$$\mathbf{M} = \begin{pmatrix} 0 & 0 \\ 1 & 0 \end{pmatrix}.$$

$$\mathbf{M}^2 = \begin{pmatrix} 0 & 0 \\ 1 & 0 \end{pmatrix} \begin{pmatrix} 0 & 0 \\ 1 & 0 \end{pmatrix} = \begin{pmatrix} 0 & 0 \\ 0 & 0 \end{pmatrix}$$

node
a point at which two or more curves meet

nonagon
a nine-sided polygon

non-Euclidean
not assuming or in accordance with all of Euclid's postulates of
geometry

nontrivial

of or being a solution to an equation in mathematics in which at least one unknown value is not equal to zero – compare TRIVIAL SOLUTION

> **Example** A nontrivial solution of the equations
> $$x + y + z = 0$$
> $$x - y = 0$$
> $$2y + z = 0$$
> is $x = 1$, $y = 1$, and $z = -2$

normal (*noun*)

a line, plane, or vector that is perpendicular to another line, plane, or curve

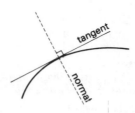

normal (*adjective*)

1 *in geometry* perpendicular; *esp* perpendicular to a tangent. NB the vector

$$\begin{pmatrix} a \\ b \\ c \end{pmatrix}$$

is normal to the plane $ax + by + cz = 0$.

2 *in probability & statistics* of, involving, or being a normal curve or normal distribution ⟨~ *approximation to the binomial distribution*⟩

normal curve

the symmetrical bell-shaped curve of a normal distribution

normal distribution

a probability density function that approximates the distribution of many random variables and has a symmetrical bell-shaped graph

normalize *or* normalise

to multiply a vector by a factor which makes it a unit vector

null hypothesis

a statistical hypothesis to be tested and accepted or rejected in favour of an alternative; *specif* the hypothesis that an observed difference (eg between the means of two samples) is due to chance alone

number line
the representation of real numbers as points on a line

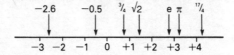

number theory
a branch of mathematics dealing with the integers and their properties

numeral
a conventional symbol that represents a natural number or zero
   Arabic numerals: 0, 1, 2, 3, ... 9
   Roman numerals: I, V, X, L, C, D, M etc

numerator
the part of a fraction that is above the line (eg the numerator of ³⁄₇ is 3) – compare DENOMINATOR

numerical analysis
a branch of mathematics concerned with substituting sets of numerical data into mathematical models to obtain approximations to the solutions of equations; *esp* the application and theory of such methods involving computer programs

oblong
rectangular but not square

obtuse
of an angle greater than 90° but less than 180°

octagon
a polygon having eight sides

octahedron
a solid having eight faces. A regular octahedron has eight
equilateral triangular faces.

octal
of, being, or belonging to a system of counting having eight as its
base (eg the octal number 237_8 is equal to the decimal number
$2 \times 8^2 + 3 \times 8 + 7 = 159$)

odd
not divisible by two without leaving a remainder (eg 1, 3, and 5
are odd numbers)

odd function
a function f(x) such that f($-a$) = $-$f(a)

   Examples
     (a) f(x) = x, x^3, x^5 ...
     (b) f(x) = $\sin x, \tan x$

  – compare EVEN FUNCTION

ogive
a cumulative frequency graph

one-sided surface
– see MÖBIUS STRIP

one-tailed test *or* **one-tail test**
a statistical test of a hypothesis, in which all the possible values of
the test statistic (statistic used in testing a hypothesis) that would
lead to rejection of the hypothesis are either greater than a given
value or less than some other given value, but not both – compare
TWO-TAILED TEST

one-to-one function
a function which maps *one* member of the domain onto *one*

member of the co-domain *and vice versa*

$$(eg\ x \mapsto x^3,\ but\ not\ x \mapsto x^2)$$

open interval

an interval which does not include the end points (eg $3<x<5$)

operation

a mathematical process carried out to derive one expression from others according to a rule (eg $+\ -\ \times\ \div$ are binary operations; $\sqrt{}$ is a unary operation)

operational research

the application of scientific, esp mathematical, methods to the study and analysis of problems involving complex systems (eg business management, economic planning, and the waging of war)

opposite

1 of or being the side of a triangle that faces a given angle. In a triangle ABC, the side BC is opposite the angle A

2 – see VERTICALLY OPPOSITE ANGLES

order

1 a sequential arrangement of mathematical elements

2 *of a derivative* the number of times differentiation is applied successively (eg d^3y/dx^3 is the third order derivative of y with respect to x)

3 *of a differential equation* the order of the highest derivative present in the equation

4 *of a matrix* the number of rows and columns. A matrix with m rows and n columns has the order $m \times n$

5 the number of elements in a finite mathematical group

ordered pair

a pair of real numbers arranged according to some rule (eg (x,y) the x- and y- coordinates of a point)

ordinal number

a number designating the place (eg first, second, or third) occupied by an item in a sequence

ordinate

the coordinate of a point in a Cartesian coordinate system obtained by measuring parallel to the y-axis; the y-coordinate of a point in a plane – compare ABSCISSA

origin

the point of intersection of coordinate axes

orthocentre
the point of intersection of the
altitudes of a triangle

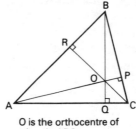

O is the orthocentre of
triangle ABC

orthogonal
mutually perpendicular

orthogonal matrix
a square matrix **M** such that $\mathbf{MM}^T = \mathbf{I}$ (where $\mathbf{M}^T$ is the transpose
of **M**, and **I** is the identity matrix)

Example $\mathbf{M} = \begin{pmatrix} 0{\cdot}6 & -0{\cdot}8 \\ 0{\cdot}8 & 0{\cdot}6 \end{pmatrix}$ $\quad \mathbf{M}^T = \begin{pmatrix} 0{\cdot}6 & 0{\cdot}8 \\ -0{\cdot}8 & 0{\cdot}6 \end{pmatrix}$

$$\mathbf{MM}^T = \begin{pmatrix} 0{\cdot}6 & -0{\cdot}8 \\ 0{\cdot}8 & 0{\cdot}6 \end{pmatrix} \begin{pmatrix} 0{\cdot}6 & 0{\cdot}8 \\ -0{\cdot}8 & 0{\cdot}6 \end{pmatrix} = \begin{pmatrix} 1 & 0 \\ 0 & 1 \end{pmatrix}$$

orthonormal
being normalized and orthogonal, eg the vectors

$$\begin{pmatrix} 0{\cdot}6 \\ 0{\cdot}8 \end{pmatrix} \text{ and } \begin{pmatrix} -0{\cdot}8 \\ 0{\cdot}6 \end{pmatrix}$$

Osborn's rule
a rule stating that an identity concerning trigonometrical functions
may be changed into the corresponding identity for hyperbolic
functions provided the signs are changed wherever there is a
product (or implied product) of two sines, eg
$$\cos (A + B) = \cos A \cos B - \sin A \sin B$$
becomes $\cosh (A + B) = \cosh A \cosh B + \sinh A \sinh B$

oscillate
to vary above and below a midpoint or average value

ounce (*abbr* **oz**)
an imperial unit of mass equal to $\frac{1}{16}$ of a pound

P

parabola
a two-dimensional curve generated by a point P moving so that its distance from a fixed point S (the focus) is equal to its distance from a fixed line l (the directrix); the intersection of a cone with a plane parallel to a generator – compare CONIC SECTION, ELLIPSE, HYPERBOLA

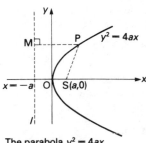

The parabola $y^2 = 4ax$

paraboloid
a surface or solid generated by the rotation of a parabola about its axis of symmetry

parallel
extending in the same direction and always being the same distance apart. When a transversal intersects parallel lines, the corresponding angles are equal.

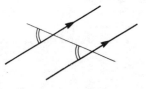

parallelepiped
a polyhedron whose six faces are parallelograms

parallelogram
a quadrilateral with opposite sides parallel and equal

parameter
an arbitrary real number used as the independent variable in expressions which give the coordinates of a point on a curve (or surface)

Examples

curve	parametric form
parabola	$(at^2, 2at)$
circle	$(a\cos t, a\sin t)$
ellipse	$(a\cos t, b\sin t)$
hyperbola	$(a\sec t, b\tan t)$
rectangular hyperbola	$(ct, c/t)$

parity
the property of an integer with respect to being odd or even (eg 3 and 7 have the same parity)

partial derivative
the derivative of a mathematical function of several variable quantities with respect to any one of them and with the remaining variables treated as constants

partial differential equation
a differential equation containing at least one partial derivative

partial fraction
any of the simpler fractions that when summed are equivalent to a quotient of two or more polynomials

> **Example**
> Express $\dfrac{2x}{(x + 1)(x - 1)}$ in partial fractions.
>
> $$\frac{2x}{(x + 1)(x - 1)} = \frac{1}{x - 1} + \frac{1}{x + 1}$$

Pascal's triangle
a triangular array of numbers consisting of rows that are the coefficients of the expansion of $(a + b)^n$ for successive integers n beginning with 0, and that is used in probability theory.

$$
\begin{array}{ccccccccccc}
 & & & & 1 & & 1 & & & & \\
 & & & 1 & & 2 & & 1 & & & \\
 & & 1 & & 3 & & 3 & & 1 & & \\
 & 1 & & 4 & & 6 & & 4 & & 1 & \\
1 & & 5 & & 10 & & 10 & & 5 & & 1 \\
\end{array}
$$

The number in each row is the sum of the two numbers just above it.

pentagon
a five-sided polygon

percentage
a proportion expressed as a number out of a hundred
(eg $3/5 = 60/100 = 60\%$; $25/80 = 0.3125 = 31.25\%$)

percentile
any of 99 values in a frequency distribution that divides it into 100 parts (intervals), each containing 1 per cent of the items under consideration

perfect number

an integer that is equal to the sum of all the integers by which it can be divided without leaving a remainder, including 1 but excluding itself (eg $28 = 1 + 2 + 4 + 7 + 14$)

perfect square

the square of an integer (eg 1, 4, 9, 16, 25, ...)

perimeter

(the length of) the boundary of a closed plane figure (eg a square or circle)

period

1 the time or interval of time that elapses before a cyclic motion or phenomenon begins to repeat itself

2 a number that does not change the value of a periodic function when added to the independent variable; *esp* the smallest such number

3 *of an element, a, of a group* the smallest integer n, such that $a^n = e$, the identity element

periodic function

a mathematical function (eg sine or cosine) whose values recur at regular intervals

Examples

function	period (in radians)
sin t	2π
cos t	2π
tan t	π
sin nt	$2\pi/n$

permutation

an ordered arrangement of a given set of distinct objects (eg the permutations of ABC are ABC ACB BCA BAC CAB CBA). The number of permutations of n objects is $n!$;

the number of permutations of r objects selected from n distinct objects is

$$^n\mathrm{P}_r = \frac{n!}{(n - r)!}$$

– compare COMBINATION

perpendicular

being at right angles to a given line or plane

NB If lines $y = m_1x + c_1$ and $y = m_2x + c_2$ are perpendicular, then $m_1m_2 = -1$.

If vectors **a** and **b** are perpendicular, then the scalar product **a.b** is zero.

pi (*symbol* π)
the ratio of the circumference of a circle to its diameter – see
GREGORY'S SERIES
NB π cannot be expressed as a rational number, but 3·142 and
22/7 are often used as approximations. A very ancient value for π
can be found in the Bible, I Kings Ch 7, v 23 (Authorized
Version).

pie chart
a diagramatic representation of statistics, in which the areas of
sectors of a circle are proportional to the data

place value
the value of the position of a digit in a numeral (eg in 425 the
location of the digit 2 has a place value of 10)

plane
a flat surface. The equation of a plane in Cartesian coordinates has
the form $ax + by + cz = d$; the plane is perpendicular to the
vector $a\mathbf{i} + b\mathbf{j} + c\mathbf{k}$.

plane geometry
a branch of elementary geometry that deals with two-dimensional
figures

Platonic solid
REGULAR SOLID

point
1 a geometric element that has a position in space but no size
2 a geometric element determined by an ordered set of
coordinates ⟨*a* ~ *on a graph*⟩

point of inflection
– see INFLECTION

Poisson distribution
a probability distribution that is often used to model the number
of outcomes of discrete events (eg traffic accidents or atomic
disintegrations) that occur in a continuum, and in which the
probability of the event happening is small but the number of
times at which the event can happen is large, and which can be
used as an approximation to the binomial distribution

polar coordinate
either of two numbers that
locate a point in a plane by its
distance *r* along a line from a
fixed point (the pole) and the
angle θ this line makes with a
fixed line – compare
CARTESIAN COORDINATE

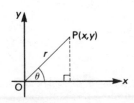

pole
1 either extremity of an axis of a sphere or of a body, esp the earth, resembling a sphere
2 the fixed point in a system of polar coordinates that serves as the origin

polygon
a closed two-dimensional geometric figure bounded by straight lines

polyhedron
a solid three-dimensional geometric figure bounded by plane faces
– see EULER'S RELATION

polynomial
a sum of two or more algebraic terms each of which consists of a constant multiplied by a variable raised to a nonnegative integral power (eg $2 + 3x + 4x^2 - 7x^3 + x^4$)

population
a set (eg of individual people or items) from which samples are taken for statistical measurement

population density
– see DENSITY

position vector
the vector $\overrightarrow{OP}$ (= **p**) which defines the displacement from the origin to point P (eg the position vector of the point P (3,4,5) is **p** = 3**i** + 4**j** + 5**k**)

positive
1 numerically greater than zero ⟨*+2 is a ~ integer*⟩
2 extending or generated in a direction chosen to be positive; *esp* of or being that part of the *x*-axis that extends to the right of the origin
3 *of an angle* measured in an anticlockwise direction

postulate
a statement that is accepted without proof

potential energy
energy due to position. The potential energy of a weight w, raised to a height h is wh. The potential energy of an elastic string, natural length l, modulus of elasticity λ, extension x, is

$$\frac{1}{2} \frac{\lambda}{l} x^2$$

– compare KINETIC ENERGY

pound (*abbr* **lb**)
an imperial unit of mass equal to approx 0·45kg

power

1 the number of times a given number is (to be) multiplied by itself ⟨*the cube is the third ~ of a number*⟩

2 the amount of work done or energy emitted or transferred per unit of time; force × velocity – see WATT

power series

an infinite series whose terms are successive integral powers of a variable multiplied by constants and that takes the form
$a + bx + cx^2 + dx^3 + ex^4 + ...$

prime

having no factors except the number one. A prime number is any positive integer except the number one (1) that has no factor except itself and 1 (eg 2, 3, 5, 7, 11, 13, 17, ...) – see CO-PRIME

primitive

INDEFINITE INTEGRAL (eg x^3 is a primitive of $3x^2$)

principal diagonal

the diagonal in a square matrix that runs from upper left to lower right – called also LEADING DIAGONAL

prism

a three-dimensional polyhedron having two equal ends which are polygons lying in parallel planes connected by sides which are parallelograms

probability

1 a measure of the likelihood that a given event will occur, usu expressed as a fraction between 0 (impossible) and 1 (certain).
a If a 'successful' outcome of an experiment can occur in s ways, and n is the total number of equally likely outcomes, then the probability of a 'success' is s/n (eg the probability of scoring a six when a fair die is rolled is 1/6).

b If a 'successful' event occurs s times in n trials (where n is large), then the probability of a 'success' is

$$\lim_{n \to \infty} (s/n).$$

2 a branch of mathematics concerned with the study of probability

probability density function

a function of a continuous random variable whose integral over an interval gives the probability that its value will fall within the interval

probability function

a function of a discrete random variable that gives the probability that a specified value will occur

product

the result of multiplying together two or more numbers or expressions

product rule

a rule stating that if u and v are functions of x, then

$$\frac{d}{dx}(uv) = u\frac{dv}{dx} + v\frac{du}{dx}$$

progression

a sequence of numbers in which each term is related to its predecessor by a uniform law – compare ARITHMETIC PROGRESSION, GEOMETRIC PROGRESSION, HARMONIC PROGRESSION, FIBONACCI SEQUENCE

projection

of a line onto a plane the line formed by dropping perpendiculars from two points (M and N) on a given line onto a plane. If the feet of the perpendiculars are P and Q, then the line PQ is the projection of MN onto the plane.

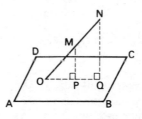

proof

the process of establishing the truth or validity of a statement

proper

being a mathematical subset that does not contain all the elements of the inclusive set from which it is derived. For example, 5 is a proper factor of 35 (but 1 and 35 are not); {a,b} is a proper subset of {a,b,c,d} (but the empty set and {a,b,c,d} are not).

proper fraction

1 *in arithmetic* a fraction in which the numerator is less than the denominator (eg ³⁄₇)

2 *in algebra* a fraction on which the numerator is a polynomial of lower degree than the polynomial in the denominator (eg

$$\frac{x^2 + x + 3}{2x^3 - 4x + 1}$$

– compare IMPROPER FRACTION

proportion

1 the relation of one part to another or to the whole with respect to magnitude, quantity, or degree; a ratio

proposition

2 a statement of equality of ratios. Two sets of numbers are said to be in proportion when corresponding elements are in the same ratio (eg {2, 3, 7, 10} and {6, 9, 21, 30} – see DIRECT PROPORTION, INVERSE PROPORTION

proposition
a formal mathematical statement to be proved

protractor
an instrument that is used for marking out or measuring angles

prove
to establish the truth of a theorem or proposition

pyramid
a three-dimensional figure having a base that is a polygon and other faces which are triangles with a common vertex

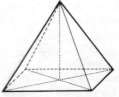

A pyramid on a square base

Pythagoras (570–501 BC) Greek geometer

Pythagoras' theorem
a theorem in geometry: the square of the length of the hypotenuse of a right-angled triangle equals the sum of the squares of the lengths of the other two sides

Pythagorean triple
a set of three integers (a,b,c) such that $a^2 + b^2 = c^2$, eg (3, 4, 5) (5, 12, 13) (8, 15, 17)

Q

quadrant
1 the area of one quarter of a circle that is bounded by a quadrant and two radii at right angles to one another
2 any of the four parts into which a plane is divided by two axes lying at right angles to each other in that plane

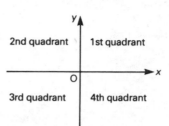

quadratic (*adjective*)
of or involving (mathematical terms of) the second degree

quadratic (*noun*)
an algebraic expression including terms of the second degree (but no higher degree)

quadratic equation
an equation of the form $ax^2 + bx + c = 0$

quadratic formula
a formula for solving a quadratic equation

$$x = \frac{-b \pm \sqrt{(b^2 - 4ac)}}{2a}$$

– see DISCRIMINANT

quadratic function
a function of the form $f(x) = ax^2 + bx + c$. Its graph is parabolic.

quadratic polynomial
a polynomial of the form $ax^2 + bx + c$

quadrature
the process of finding a square equal in area to a given surface or figure

quadrilateral
a four-sided polygon – see CYCLIC QUADRILATERAL, KITE, PARALLELOGRAM, SQUARE, RECTANGLE, RHOMBUS, TRAPEZIUM

quadruple
to make or become four times as great

quartic
an algebraic expression including terms of the fourth degree

quartile

quartile
any of three values in a frequency distribution that divide it into four intervals, each containing one quarter of the data. The first quartile (*lower quartile*) corresponds to the 25th percentile and is the value below which 25 per cent of the data lies, the middle or second quartile is the median, and the third or highest quartile (*upper quartile*) corresponds to the 75th percentile.

quaternion
a generalized complex number that contains one real part and three imaginary parts

quintic equation
an equation of the fifth degree. For centuries, many famous mathematicians (including Abel and Galois) tried to find an algebraic solution to the general quintic equation, but failed. Hermite in 1842 proved that this is impossible.

quotient
the result of the division of one number or mathematical expression by another (eg in $15 \div 3 = 5$, the quotient is 5)

quotient rule
a rule stating that if u and v are functions of x, then

$$\frac{d}{dx}\left(\frac{u}{v}\right) = \frac{v\frac{du}{dx} - u\frac{dv}{dx}}{v^2}$$

R

radial
relating to or directed along a radius ⟨~ *acceleration*⟩

radian
a unit of angular measurement that is equal to the angle between two radii of a circle that subtend an arc of the circumference equal to the length of the radius; approx 57·3° – see DEGREE

radical sign
the sign $\sqrt{}$ placed before a mathematical expression to denote that the square root is to be calculated, or some other root corresponding to an index number placed before the sign (eg $_3\sqrt{}$)

radius (*plural* **radii**)
(the length of) a straight line extending from the centre of a circle or sphere to the circumference or surface – compare DIAMETER

radius of curvature
the reciprocal of the curvature of a curve

radius vector
the position vector of a variable point P in polar coordinates

random
relating to, having, or being statistical elements or events with an ungoverned or unpredictable outcome

randomize *or* **randomise**
to arrange (eg samples or experimental treatments) so as to simulate a chance distribution, reduce interference by irrelevant variables, and yield unbiased statistical data

random variable
a statistical variable that can take on a defined range of values which are governed by a probability distribution (eg the number of spots showing if two dice are thrown is a random variable)

random walk
a statistical process consisting of a sequence of steps each of whose characteristics (eg magnitude and direction) are determined by chance

range
1 the difference between the least and greatest values of a mathematical function, sequence, or series – see INTERQUARTILE RANGE

2 the set of images of a function – compare DOMAIN

Example Given the domain $\{-3, -2, -1, 0, +1, +2, +3\}$ the range of

(a) $x \mapsto 2x$ is $\{-6, -4, -1, 0, +2, +4, +6\}$

(b) $x \mapsto x^2$ is $\{0, +1, +4, +9\}$

rank

the order according to some statistical characteristic (eg a score in a test)

rank correlation

a measure of correlation depending on rank

rate of change

the rate of change of a variable, x, with respect to another, t, is given by dx/dt

ratio

the relationship between two numbers (or quantities measured in the same units) obtained by dividing one by the other; written in the form $a{:}b$ or a/b.

Examples

(a) $60{:}20 = 3{:}1$, $35{:}14 = 5{:}2 = 2{\cdot}5{:}1$, £3:20p $= 300{:}20 = 15{:}1$.

(b) Share £1000 in the ratio 3:2.

$$\text{Larger share} = \frac{3}{5} \times £1000 = £600;$$

$$\text{smaller share} = \frac{2}{5} \times £1000 = £400.$$

(c) The point P divides the line segment AB in the ratio 3:7. If AP = 20cm, the lengths of AP and PB are:

$$\text{AP} = \frac{3}{10} \times 20\text{cm} = 6\text{cm};$$

$$\text{PB} = \frac{7}{10} \times 20\text{cm} = 14\text{cm}.$$

rational function

1 POLYNOMIAL

2 a function that is the quotient of two polynomials

$$\left(\text{eg } \frac{x + 1}{x^2 + 3x + 5}\right)$$

rationalize *or* **rationalise**

to free the denominator of a fraction from irrational numbers

Examples

(a) $\dfrac{6}{\sqrt{2}} = \dfrac{6}{\sqrt{2}} \times \dfrac{\sqrt{2}}{\sqrt{2}} = \dfrac{6 \times \sqrt{2}}{2} = 3\sqrt{2}$

(b) $\dfrac{1}{2 - \sqrt{3}} = \dfrac{1}{2 - \sqrt{3}} \times \dfrac{2 + \sqrt{3}}{2 + \sqrt{3}} = \dfrac{2 + \sqrt{3}}{4 - 3} = 2 + \sqrt{3}$

rational number
a number (eg $\frac{3}{7}$, $\frac{5}{8}$, $1\frac{1}{3}$) of the form a/b, where a and b are
integers and $b \neq 0$. (NB a rational number may be greater than 1.)
NOTATION The symbol for the set of rational numbers is $\mathbb{Q}$.
– compare REAL NUMBER, IRRATIONAL NUMBER, INTEGER, SURD

ratio theorem
a theorem stating that if P is a
point on the line segment AB
which divides AB in the ratio
$m{:}n$, then the position vector of
P is related to the postion
vectors of A and B by the
equation
$$(m + n)\mathbf{p} = n\mathbf{a} + m\mathbf{b}$$

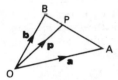

raw score
an individual's actual score in a test before any statistical
treatment or adjustment

ray
1 a thin line suggesting a ray
2 any of a group of lines diverging from a common centre
3 a straight line extending from a point in one direction only

real
belonging to the set of real numbers

real number
any of the numbers that have no imaginary parts and comprise the
rational numbers and the irrational numbers; a number that can be
represented by a point on the number line (eg -3, $-\frac{1}{2}$, 0, 0·75,
$1\frac{3}{4}$, 5, e, π, 17/4). The symbol for the set of real numbers is $\mathbb{R}$.
– compare RATIONAL NUMBER, INTEGER

real part
the part of a complex number that does not include the imaginary
number. The real part of a complex number $z = a + b\mathrm{i}$ (written

$\mathrm{Re}(z)$) is a (eg if $z = 4 + 3\mathrm{i}$, $\mathrm{Re}(z) = 4$) – compare IMAGINARY PART

reciprocal
1 either of any two numbers (eg ⅔, 3⁄2) which when multiplied together give one
2 the number obtained by dividing one by a particular number (eg the reciprocal of 10 is 0·1)

rectangle
a parallelogram whose angles are right angles

rectangular
1 shaped like a rectangle ⟨$a \sim area$⟩
2 crossing, lying, or meeting at a right angle ⟨$\sim axes$⟩
3 having edges, surfaces, or faces that meet at right angles; having faces or surfaces shaped like rectangles

rectangular coordinate
CARTESIAN COORDINATE

rectangular hyperbola
a hyperbola whose asymptotes
are perpendicular; *esp* the
hyperbola $xy = c^2$

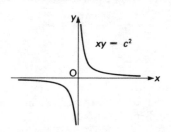

rectify
to determine the length of an arc or curve

rectilinear
1 moving in, being in, or forming a straight line ⟨$\sim motion$⟩
2 characterized by straight lines

recurring decimal
a decimal in which a particular digit or sequence of digits repeats itself indefinitely at some stage after the decimal point
(eg $0 \cdot \dot{3} = 0.333 \ldots = $ ⅓; $0 \cdot \dot{1}\dot{8} = 0.181818 \ldots = $ 2⁄11)

recursion
the repeated application of a particular mathematical procedure to the previous result to determine either a sequence of numbers or a more accurate approximation to a square root, fraction, etc – see also ITERATIVE

recursive
of or involving mathematical recursion

reduce
1 to change the denominations or form of without changing the value (eg the fraction $^{14}/_{21}$ reduced to its lowest terms is $^2/_3$)
2 to construct a geometrical figure similar to but smaller than a given figure

reductio ad absurdum
reduction to absurdity; an indirect method of proof in which it is shown that the opposite of the hypothesis leads to a contradiction

redundancy *also* **redundance**
the part of a statement that can be eliminated without loss of essential information

reflection
a transformation of a geometrical figure with respect to an axis, producing a mirror image of the figure – compare ROTATION, TRANSLATION, SHEAR, ENLARGEMENT

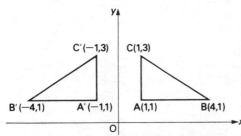

Triangle A'B'C' is the reflection of triangle ABC in the *y*-axis

reflex
of an angle greater than 180° but less than 360°

region
a set of points S in two (or three) dimensions such that any pair of points in S can be joined by a continuous curve, lying wholly inside set S

regress
to tend to approach or revert to a mean

regression
(the statistical analysis of) a mathematical relationship between two or more correlated variables that is often determined directly from observed data and is used esp to predict values of one variable when given values of the other

regular polygon

a polygon with all its sides and angles equal

regular solid

a polyhedron with all its faces equal. There are five regular polyhedrons: tetrahedron, cube, octahedron, dodecahedron, and icosahedron

relation

an aspect or quality that connects two or more things; *specif* a property (eg one expressed by 'is equal to' or 'is less than') that holds between an ordered pair of objects or numbers

relative

expressed as the ratio of the specified quantity (eg an error in measuring) to the total magnitude (eg the value of a measured quantity) or to the mean of all the quantities involved $\langle \sim error \rangle$

relative frequency

in statistics the ratio of 'successful' outcomes of an experiment, to the total number of trials. For example, if a die is rolled 100 times and 17 sixes are recorded, the relative frequency of the score six is 17/100. – see PROBABILITY

relatively prime

CO-PRIME

remainder

1 *in arithmetic* the (whole) number which is left when one integer is divided by another (the remainder must be less than the divisor)

Example $17 = 3 \times 5 + 2$, so the remainder when 17 is divided by 3 is 2.

2 *in algebra* the polynomial which is left when one polynomial is divided by another (the remainder must be of lower degree than the divisor)

Examples

(a) $x^3 + 3x^2 + x + 5 = (x + 3)(x^2 + 1) + 2$, so the remainder when $x^3 + 3x^2 + x + 5$ is divided by $(x + 3)$ is 2.

(b) When $x^5 + 4x^2 + x - 1$ is divided by $x^2 - 1$, we obtain $x^5 + 4x^2 + x - 1 = (x^2 - 1)(x^3 + x + 4) + (2x + 3)$. Here the remainder is $(2x + 3)$.

remainder theorem

a theorem stating that when a polynomial $P(x)$ is divided by $(x - a)$ the remainder is $P(a)$

Example If $P(x) = x^3 + 3x^2 + x + 5$, the remainder when $P(x)$ is divided by $(x + 3)$ is
$P(-3) = -27 + 27 - 3 + 5 = 2$

– see FACTOR THEOREM

repeat

to occur more than once (eg the equation $x^2 - 6x + 9 = 0$ has a repeated root, $x = 3$)

repeating decimal

RECURRING DECIMAL

residue

the remainder after subtracting a multiple of a modulus from an integer ⟨*2 and 7 are ~s of 12 modulo 5*⟩; *esp* the smallest nonnegative instance of such a residue

resolve

to express as the sum of two or more components; *esp* to express a vector as the sum of two or more components in assigned directions

resolved part

the magnitude of a component of a vector. If **n** is a unit vector, the resolved part of a vector **F** in the direction of **n**, is the scalar product of **F** and **n**, ie **F.n**.

result

something obtained by calculation or investigation

resultant (*adjective*)

derived or resulting from something else, esp as the total effect of many causes

resultant (*noun*)

the single vector that is the sum of a given set of vectors

revolution

1 a progressive motion of a body round a centre or axis so that any line of the body remains parallel to and returns to its initial position; ROTATION 1 – compare SOLID OF REVOLUTION
2 motion of any figure about a centre or axis
3 one complete turn

revolve

1 to cause to turn round (as if) on an axis; rotate
2 to move in a curved path round a centre or axis

rhombus

an equilateral parallelogram – compare KITE, PARALLELOGRAM, SQUARE

Riemann integral

DEFINITE INTEGRAL

right

having the vertex directly above the centre of the base ⟨*a ~ cone*⟩ ⟨*a ~ pyramid*⟩.

right angle
one quarter of a revolution; an angle of 90° ($\pi/2$ radians)

right cone
– see CONE 2

right-handed set
of axes (or vectors) a set of mutually perpendicular axes (usu designated x-, y- and z-) whose directions can be demonstrated by stretching out the thumb (x), first finger (y) and middle finger (z) of the right hand.

In the diagram, the x-axis comes *out* of the page, at right-angles to the y- and z- axes. In a *left*-handed set, the direction of the x-axis would be reversed.

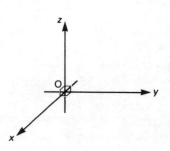

rigour
strict mathematical or logical precision

ring
a mathematical group that is closed under two binary operations (eg addition and multiplication) in which the first operation is commutative and the second operation is associative and distributive relative to the first

Rolle's theorem
a theorem in calculus: if a curve is continuous, crosses the x-axis at two points, and has a tangent at every point on the curve between those two points, the tangent to the curve is parallel to the x-axis somewhere between the two points

roman numeral (*often cap* R)
a numeral in a system of notation based on the ancient Roman system using the symbols I for "one", V for "five", X for "ten", L for "fifty", C for "hundred", D for "five hundred", and M for "thousand"

root
1 a number that, when multiplied by itself a stated number of times, gives another stated number (eg 2 is the fourth root of 16)
2 a number that, when substituted for the variable, satisfies a particular equation (eg 3 is a root when $x^2 - 5x + 6 = 0$) – see SOLUTION SET

root-mean-square

the square root of the arithmetic mean of the squares of a set of numbers

root-mean-square deviation

STANDARD DEVIATION

root of unity

an nth root of unity is one of the n complex numbers of the form:

$$\cos\left(\frac{2k\pi}{n}\right) + i \sin\left(\frac{2k\pi}{n}\right)$$

where $k = 0, 1, 2, \ldots (n-1)$.

– see DE MOIVRE'S THEOREM

rotate

to (cause to) turn or revolve round an axis or a centre; *specif* to (cause to) move in such a way that all points turn through the same angle about a single fixed point

rotation

1 a transformation of a geometrical figure in which the figure rotates round a fixed point – compare REFLECTION, TRANSLATION, ENLARGEMENT, SHEAR

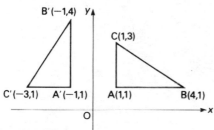

Triangle A'B'C' is the image of triangle ABC under a rotation of 90° about the origin.

2 one complete turn; the angular displacement required to return a rotating body or figure to its original orientation

rotational symmetry

if a plane figure can be rotated about a point (or a solid about an axis) to a new position, in which its appearance is unchanged, it is said to have rotational symmetry about that point (or axis); if there are n distinct positions, it is said to have rotational symmetry of order n (eg an equilateral triangle has rotational symmetry of order 3; a pyramid on a square base has rotational symmetry of order 4)

round

to correct a number to a specific degree of accuracy (eg 11·3572 rounded off to three decimal places becomes 11·357; £15·95 becomes £16 when rounded to the nearest pound)

ruler

a smooth-edged strip of plastic, wood, etc that is usu marked off in units (eg centimetres) and is used for drawing lines or measuring

S

saddle point
a point at which the value of a function of two mathematical variable quantities is a maximum with respect to one quantity and a minimum with respect to the other

sample (*noun*)
a subset of a statistical population whose properties are studied to gain information about the whole

sample (verb)
to take a sample from (a population)

sample space
a set consisting of all the possible outcomes of a statistical experiment

scalar
1 a real number, esp one used to enlarge a vector
2 **scalar, scalar quantity** a quantity (eg mass or time) that has magnitude but no direction – compare VECTOR

scalar product
a real number obtained by multiplying together the lengths of two vectors and the cosine of the angle between them – compare VECTOR PRODUCT

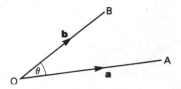

In the diagram, the scalar product of the vectors **a** and **b** (written **a.b**) is $ab \cos \theta$.
If **a** and **b** are two vectors such that
$$\mathbf{a} = x_1\mathbf{i} + y_1\mathbf{j} + z_1\mathbf{k}$$
$$\mathbf{b} = x_2\mathbf{i} + y_2\mathbf{j} + z_2\mathbf{k}$$
then their scalar product is given by
$$\mathbf{a}.\mathbf{b} = x_1x_2 + y_1y_2 + z_1z_2$$

scalar quantity
SCALAR 2

scalar triple product
a scalar quantity which is the product of three vectors **a**, **b** and **c** and is given by $\mathbf{a}.(\mathbf{b} \wedge \mathbf{c})$. It is equal to the volume of the

scale

parallelepiped whose edges are parallel to OA, OB and OC. If

$\mathbf{a} = x_1\mathbf{i} + y_1\mathbf{j} + z_1\mathbf{k}$
$\mathbf{b} = x_2\mathbf{i} + y_2\mathbf{j} + z_2\mathbf{k}$
$\mathbf{c} = x_3\mathbf{i} + y_3\mathbf{j} + z_3\mathbf{k}$

then

$$\mathbf{a} \cdot (\mathbf{b} \wedge \mathbf{c}) = \begin{vmatrix} x_1 & y_1 & z_1 \\ x_2 & y_2 & z_2 \\ x_3 & y_3 & z_3 \end{vmatrix}$$

– compare VECTOR TRIPLE PRODUCT

scale
a proportion between two sets of dimensions (eg between two similar triangles)

scale factor
the factor by which the dimensions of a figure must be multiplied when it is enlarged

scalene
of a triangle having the three sides of unequal length

scatter
DISPERSION

scatter diagram
a two-dimensional graph consisting of points whose positions represent values of two variables and illustrating the correlation of the variables

scientific notation
a system in which numbers are expressed as products consisting of a number between 1 and 10 and a power of 10 (eg 14 000 in scientific notation is $1 \cdot 4 \times 10^4$; 0·014 is $1 \cdot 4 \times 10^{-2}$) – called also STANDARD FORM

score
1 twenty
2 a group of twenty things

sea mile
NAUTICAL MILE

secant
the mathematical function that is the reciprocal of cosine

second
1 a 60th part of a minute of time or of a minute of angular measurement
2 the basic SI unit of time, equal to the duration of 9 192 631 770 oscillations of the radiation corresponding to the transition between two specific energy levels of the caesium isotope $_{55}Cs^{133}$

section

1 a shape or area as it would appear if a solid form were cut through by one plane

2 the plane figure resulting from the cutting through of a solid form by one plane (eg conic section) – see also GOLDEN SECTION

sector

a part of a circle bounded by two radii and the portion of the circumference between them. The smaller part is called the minor sector and the larger part the major sector – compare SEGMENT 2

In the diagram the area of the minor sector AOB = $\frac{1}{2}r^2\theta$ (θ in radians)

$$= \left(\frac{\theta}{360}\right)\pi r^2 \quad (\theta \text{ in degrees})$$

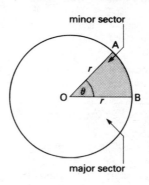

minor sector

major sector

segment

1 a portion of a geometrical figure

2 either of the two regions into which a circle is divided by a chord. The smaller region is called the minor segment and the larger region the major segment – compare SECTOR

3 a part of a sphere cut off by a plane or included between two parallel planes

4 the part of a line between two points

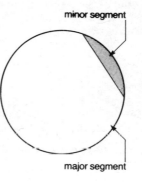

minor segment

major segment

semi-

precisely half of

semicircle

a half of a circle

sense

either of two opposite directions of motion, esp rotation

separation of variables

a method for solving (some) first order differential equations, in which the two variables are separated on either side of the equals sign

sequence

an ordered set of numbers constructed according to some rule (eg 1, 3, 5, 7, ...; 2, 4, 8, 16, 32, ...) – see FIBONACCI SEQUENCE

series

the sum of the terms of a mathematical sequence (eg $1 + 3 + 5 + 7 + ... + 99$; $2 + 4 + 8 + 16 + ... + 1024$) – see GEOMETRIC PROGRESSION, ARITHMETIC PROGRESSION, MACLAURIN'S SERIES, TAYLOR'S SERIES

set

a collection of distinct and usu mathematical objects (eg the set of even numbers less than twenty = {2, 4, 6, 8, 10, 12, 14, 16, 18})

sexagesimal

of or based on the number 60

sextuple

having six units or members

shear

a transformation of a geometrical figure in which the points on a line (in 2 dimensions) or a plane (in 3 dimensions) remain fixed, while all the other points move parallel to the fixed line (or plane), the amount of the displacement being proportional to the distance from the fixed line (or plane) – compare TRANSLATION, ENLARGEMENT, ROTATION, REFLECTION

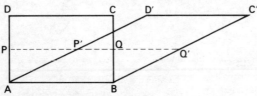

The parallelogram ABC'D' is the image, under a shear, of the rectangle ABCD

side

a boundary line of a polygon

sigma notation

the symbol

$$\sum_{r=1}^{r=n} (\text{expression})$$

is an instruction to find the sum of all the terms having the form of the given expression, for $r = 1, 2, 3, ..., n$

Examples

(a) $\sum_{r=1}^{r=10} r^2 = 1^2 + 2^2 + 3^2 + \ldots + 10^2$

(b) $\sum_{r=1}^{r=100} \left(\frac{1}{r}\right) = \frac{1}{1} + \frac{1}{2} + \frac{1}{3} + \ldots + \frac{1}{100}$

(c) $\sum_{r=0}^{r=\infty} \left(\frac{1}{2}\right)^r = 1 + \frac{1}{2} + \frac{1}{4} + \frac{1}{8} + \ldots$

significance level
LEVEL OF SIGNIFICANCE

significant
in statistics probably caused by something other than chance

significant figures
the specified number of digits in a number, that are considered to give correct or sufficient information on its accuracy, and are read from the first nonzero digit on the left to the last nonzero digit on the right, unless a final zero expresses greater accuracy; *also* the digits in a number which are farthest to the left (but excluding those zeros which are used to mark the position of the decimal point)

Example In the numbers 207 400 and 0·002 074, the first significant figure is 2, and the second is 0. When corrected to two significant figures these numbers become 210 000 and 0·0021 respectively.

similar
of geometric figures having the same shape; in particular, similar triangles have their corresponding angles equal (and their sides in proportion)

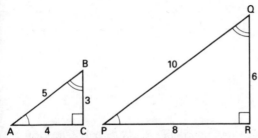

Triangle PQR is similar to triangle ABC (the scale factor is 2)

simple closed curve

a closed plane curve (eg a circle or ellipse) that does not intersect itself

simple harmonic motion (*abbr* SHM)

oscillatory motion in which a point P moves so that its acceleration is (a) proportional to its displacement, x, from a fixed point O in its path, and (b) directed towards O, ie

$$\frac{d^2x}{dt^2} = -n^2x, \text{ where } n \text{ is a constant.}$$

The period of the motion is $2\pi/n$.

simplify

an instruction to reduce a mathematical expression to a simpler form, usu one containing fewer terms

Examples (a) $2(x + y) + 5(3x - y) = 17x - 3y$

(b) $\frac{20a^2b}{5ab^2} = \frac{4a}{b}$

– see also RATIONALIZE

Simpson's rule

a numerical method for estimating the area under a curve

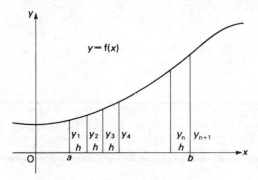

$$\int_a^b f(x)\,dx \approx \frac{h}{3}\left(y_1 + 4y_2 + 2y_3 + 4y_4 + 2y_5 + \ldots + 4y_n + y_{n+1}\right)$$

NB The number of strips must be even – compare TRAPEZIUM RULE

simultaneous

satisfied by the same values of the variables

Example the simultaneous equations $\quad 2x - 7y = 3$
$\qquad\qquad\qquad\qquad\qquad\qquad\quad x - 4y = 1$

are satisfied by $x = 5$, $y = 1$.

sine

a trigonometric function of an
angle. The sine of an acute
angle θ is the ratio of the side
opposite to the angle to the
hypotenuse in a right-angled
triangle;

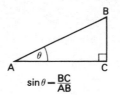

$$\sin \theta = \frac{BC}{AB}$$

$$\sin \theta = \frac{\text{opposite side}}{\text{hypotenuse}}$$

The sine of a general angle θ is
the ratio y/r in triangle OPM,
in which point P has
coordinates (x,y) and r is the
length OP. The coordinates x
and y obey the usual sign
convention; the length r is
always positive – compare
COSINE, TANGENT

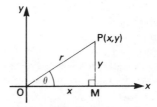

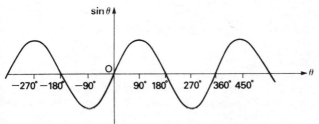

Graph of the function $\theta \mapsto \sin \theta$

sine rule

in a triangle ABC, $\qquad \dfrac{a}{\sin A} = \dfrac{b}{\sin B} = \dfrac{c}{\sin C}$

– compare COSINE RULE

Example In triangle ABC, $\angle A = 60°$, $\angle B = 40°$ and $c = 5$, find
a.

$C = 180° - (40° + 60°) = 80°$.

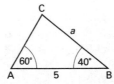

From the sine rule,

$$\frac{a}{\sin A} = \frac{c}{\sin C}$$

$$\frac{a}{\sin 60°} = \frac{5}{\sin 80°}$$

$$\therefore a = \frac{5\sin 60°}{\sin 80°}$$

= 4·40 correct to 3 significant figures.

singularity
SINGULAR POINT

singular matrix
a square matrix whose determinant is zero and which, as a consequence, has no inverse

Example

$$\mathbf{M} = \begin{pmatrix} 6 & 3 \\ 4 & 2 \end{pmatrix}$$
$$\det(\mathbf{M}) = 6 \times 2 - 4 \times 3$$
$$= 0$$

singular point
a point at which a mathematical function is undefined (eg by reason of division by zero or nonexistence of a derivative). For example, $f(x) = x^{2/3}$ has a singular point at (0,0) – see CONTINUOUS

SI units
a system of internationally agreed metric units used widely in Europe that has the metre, kilogram, second, ampere, kelvin, mole, and candela as its basic units [*Système International d'Unités*]

sketch (*noun*)
a diagram which illustrates the principal characteristics of the graph of a function (eg its stationary points and intercepts on the axes) but which is not drawn to scale

sketch (*verb*)
to make a sketch of ⟨~ *the graph of y = sin x*⟩

skew (*verb*)
1 to distort, bias
2 to cause to deviate from a true value, expected value, or symmetrical form ⟨~ed *statistical data*⟩

skew (*noun*)
1 slanting

2 *esp of a statistical distribution* more developed on one side or in one direction than another; not symmetrical

skew lines
a pair of lines (in a three-dimensional space) which are not parallel and do not meet

slant
to turn or incline from the horizontal or vertical or a correct level; slope

slant height
the length of a line from the edge of the base to the vertex of a circular-based cone

slide rule
a calculating instrument consisting in its simple form of a ruler with a central slide, both of which are graduated in such a way that the addition of lengths corresponds to the multiplication of numbers

slope
of a line GRADIENT

smooth
free from irregularities. A smooth curve is one whose gradient is continuous.

solid (*noun*)
a geometric figure (eg a cube or sphere) having three dimensions

solid (*adjective*)
having, involving, or dealing with three dimensions or with solids

solid angle
a three-dimensional spread of directions from a point (eg from the vertex of a cone or at the intersection of three planes), usu measured in steradians

solid geometry
a branch of geometry that deals with three-dimensional figures

solid of revolution
a three-dimensional figure formed by rotating the graph of a given function through 360° about an axis (usu the *x*-axis)

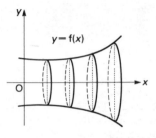

NB The cross-section perpendicular to the axis is circular.

Volume $= \int_a^b \pi y^2 \mathrm{d}x$

solution

solution
1 an action or process of solving a problem
2 an answer to a problem; an explanation; *specif* a set of values of
the variables that satisfies a mathematical equation, ie the roots of
the equation

solution set
the set of all the roots of a mathematical equation

solvable
capable of solution or of being solved

solve
to find a solution for (eg an equation)

solve a triangle
to calculate the unknown sides and angles of a triangle (or in some
cases two triangles) from given data (eg the lengths of two sides
and the size of the angle between them). Normally solving a
triangle requires the use of the sine rule and/or the cosine rule.

span
a set of vectors S is said to span a vector space V if any member of
V can be expressed as a linear combination of the members of S

speed
rate of change of distance with time, ie dx/dt, where x is the
displacement and t the time. Average speed = total distance/time
taken. Speed is a scalar quantity, the magnitude of the vector
quantity velocity.

sphere
(a volume or solid enclosed by) a surface consisting of all points at
a given distance in space from a fixed point (volume = $\frac{4}{3}\pi r^3$;
surface area = $4\pi r^2$)

spherical *also* **spheric**
1 having the form of a sphere
2 relating to or dealing with a sphere or its properties

spherical angle
the angle between two intersecting arcs of great circles of a sphere
measured by the angle formed by the tangents to the arcs at the
point of intersection

spherical coordinate
any of three coordinates that are used to locate a point in space
and that comprise the radius of the sphere on which the point lies,
the angle formed by the point, the centre of the sphere, and a
given axis of the sphere, and the angle between the plane of the
first angle and a given reference plane through the axis of the
sphere

spheroid
a mathematical figure resembling a sphere

spiral (*adjective*)
1 winding round and round a centre in ever increasing or decreasing circles; shaped like a plane spiral
2 winding round while moving up or down a central axis

spiral (*noun*)
1 the path of a point in a plane moving round a central point while continuously receding from or approaching it
2 a three-dimensional curve (eg a helix) with one or more turns about an axis

spiral of Archimedes
a plane curve that is generated by a point moving away from or towards a fixed point at a constant rate while it rotates about the fixed point at a constant rate, and that has the equation $r = a\theta$ in polar coordinates

spur
TRACE

square
1 a rectangle with all four sides equal
2 a number obtained when a number is multiplied by itself (eg the square of 5 is 25)

square bracket
either of a pair of punctuation marks [ ] used in mathematics and logic to show that a complex expression should be treated as a single unit

square matrix
a matrix with an equal number of rows and columns

square root
a number which when multiplied by itself is a given number (eg the square roots of 25 are +5 and −5).
NB the square root sign $\sqrt{}$ means the *positive* square root (eg $\sqrt{25} = +5$).

standard deviation
in statistics a measure of the spread of a population. The standard deviation of a set of n numbers $x_1, x_2, x_3, \ldots x_n$ is

$$\sqrt{\left\{ \frac{\sum(x_i - \bar{x})^2}{n} \right\}}$$

where $\bar{x}$ = the mean value of $x_1, x_2, x_3, \ldots x_n$ – compare VARIANCE

standard error
a measure used in statistics equal to the standard deviation of a set

standard form

of values divided by the square root of the total number of values

standard form *also* **standard index form**
SCIENTIFIC NOTATION

standard score
an individual test score (eg in an examination) expressed as the deviation from the average score of the group in units of standard deviation

statics
a branch of mechanics concerned with forces in equilibrium – compare DYNAMICS, KINEMATICS

stationary point
in calculus a point at which the gradient of a function is zero. This may be a local maximum, a local minimum, or a point of inflection – compare TURNING POINT

statistic
a single term or quantity in or computed from a collection of statistics; *specif* (a function used to obtain) a numerical value (eg the standard deviation or mean) used in describing and analysing statistics

statistical
of or employing the principles of statistics

statistician
1 one who compiles statistics
2 a specialist in the principles and methods of statistics

statistics
1 (*taking singular verb*) a branch of mathematics dealing with the collection, analysis, interpretation, and presentation of masses of numerical data
2 (*taking plural verb*) a collection of quantitative data

steradian
a unit of measure of solid angles that is equal to the solid angle which with its point in the centre of a sphere cuts off an area of the surface of the sphere numerically equal to the square of the radius of the sphere

Stirling's formula
a formula, $\sqrt{(2\pi n)}n^n e^{-n}$, that gives the approximate value of the factorial of a very large number

stone
an imperial unit equal to 14lb (about 6·35kg)

straight
generated by a point moving continuously in the same direction and that can be expressed by a linear equation ⟨*a* ~ *line*⟩

Student's t distribution

T DISTRIBUTION

subgroup

a subset of a mathematical group which is itself a group – see also LAGRANGE'S THEOREM

subscript

a distinguishing symbol or letter written immediately below and to the right of another character (eg x_1, x_2, x_3 ... x_n) – compare SUPERSCRIPT; called also SUFFIX

subset

a set that is included within a larger set; *esp* a mathematical set each of whose elements is also an element of a given set

NOTATION A is a subset of B is written $A \subset B$ (eg A = {a, b, c} and B = {a, b, c, d, e})

subspace

a subset of a space; *esp* one that has the properties (eg those of a vector space) of the including space

substitute

to put in the place of another, eg 4 substituted for x in the equation

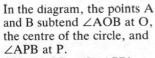

$$x^3 - 3x^2 - 10 \quad = \quad 4^3 - 3 \times 4^2 - 10$$
$$= \quad 64 - 48 - 10$$
$$= \quad 6$$

subtend

in geometry two points (esp if they are on a circle) are said to subtend an angle at a third, when they are joined to that point

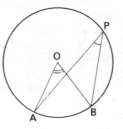

In the diagram, the points A and B subtend $\angle AOB$ at O, the centre of the circle, and $\angle APB$ at P.
(NB $\angle AOB = 2\angle APB$)

subtotal

the sum of part of a series of figures, or total of one of a group of columns of figures forming one large column

successive approximation

a method for finding an approximate value, usu the root of an equation, when an exact value is unobtainable. Starting from a first approximation, a sequence of further approximations, each one derived from its predecessor, is produced which approaches the exact value more and more closely with each successive approximation – see also ITERATION

sufficient condition

a proposition in logic or mathematics whose truth assures the truth of another proposition – compare NECESSARY CONDITION

> **Example** For a number to be divisible by 3, it is sufficient that the sum of its digits is a multiple of 3.

suffix

SUBSCRIPT

sum

1 the result of adding numbers (eg the sum of 5 and 7 is 12)
2 the mathematical limit of the sum of the first n terms of an infinite series as n increases indefinitely – see ARITHMETIC PROGRESSION, GEOMETRIC PROGRESSION

summation

1 the act or process of forming a sum; addition
2 a sum, total

superpose

to lay (eg a geometric figure) upon another so as to make all like parts coincide

superscript

a distinguishing symbol or letter written immediately above and to the right of another character (eg x^2, x^3, x^n)

supplement

1 an angle that when added to a given angle gives a total of 180°
2 a part of a circle (arc) that when added to a given arc forms a semicircle – compare COMPLEMENT 1

supplementary

of an angle or arc being either of a pair whose sum is 180° (eg 30° and 150° are supplementary angles)

supremum

the least number greater than or equal to all members of a given set of numbers

surd

an irrational number, esp a root of an integer (eg $\sqrt{3}$, $\sqrt{6}$, $\sqrt{7}$)

surface

a set of points forming the boundary of a three-dimensional region

> **Formulae for surface area**
>
> | | |
> |---|---|
> | a sphere | $4\pi r^2$ |
> | a cone (curved surface) | $\pi r l$ |
> | a cylinder (curved surface) | $2\pi r l$ |

survey (*verb*)
to determine and delineate the form, extent, and position of (eg a tract of land) by taking linear and angular measurements and by applying the principles of geometry and trigonometry

survey (*noun*)
surveying or being surveyed; *also* something surveyed

surveying
a branch of applied mathematics that deals with the determination of the area of any portion of the earth's surface, the lengths and directions of the bounding lines, and the contour of the surface, and with accurately delineating the whole on paper

surveyor
someone whose occupation is surveying land

symmetrical
1 having, involving, or exhibiting symmetry
2 having corresponding points whose connecting lines are bisected by a given point or perpendicularly bisected by a given line or plane – see also ROTATIONAL SYMMETRY

symmetric difference
the set of points belonging to one or other of two mathematical sets but not to both of them

> **Example**

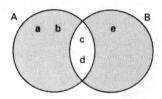

$$A = \{a, b, c, d\} \quad B = \{c, d, e\}$$
the symmetric difference $A \triangle B = \{a, b, e\}$

symmetric matrix
a matrix that is equal to its own transpose

> **Example**

$$\begin{pmatrix} 1 & 2 & 3 \\ 2 & 5 & 4 \\ 3 & 4 & 6 \end{pmatrix}$$

symmetry
the property of being symmetrical; *esp* correspondence in size, shape, and relative position of parts on opposite sides of a dividing line, centre, or other axis – compare ROTATIONAL SYMMETRY

systematic error
a statistical error that is not determined by chance but by an effect
that distorts the information in a definite direction (bias)

Système International d'Unités
– see SI UNITS

systems analysis
the analysis of an activity (eg a procedure, a business, or a
physiological function) typically by mathematical means in order
to define its goals or purposes and to discover ways of
accomplishing them most efficiently

systems analyst
a specialist in systems analysis

table

a systematic arrangement of data, usu in rows and columns, for ready reference

tangent

a trigonometric function of an angle. The tangent of an acute angle θ is the ratio of the side opposite to the angle to the adjacent side in a right-angled triangle;

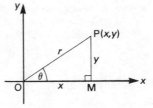

$$\tan \theta = \frac{BC}{AC}$$

$$\tan \theta = \frac{\text{opposite side}}{\text{adjacent side}}$$

The tangent of a general angle θ is the ratio y/x in triangle OPM, in which point P has coordinates (x, y) and r is the length OP. The coordinates x and y obey the usual sign convention; the length r is always positive – compare SINE, COSINE

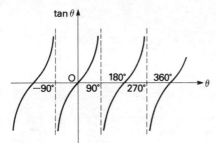

Graph of the function $\theta \mapsto \tan \theta$

tangential

1 (of the nature) of a tangent
2 acting along or lying in a tangent ⟨~ *forces*⟩

tangent line

a straight line that touches a curve

139

tangent plane

tangent plane
a plane that touches a curved surface

tangent rule
in a triangle ABC,

$$\tan\left(\frac{B - C}{2}\right) = \frac{b - c}{b + c}\cot\left(\frac{1}{2}A\right)$$

– compare SINE RULE, COSINE RULE

Taylor, Brook (1685–1731) British mathematician chiefly remembered for Taylor's series (although this was known to Gregory forty years earlier)

Taylor polynomial
a polynomial (obtained from Taylor's series) which is an approximation to a function in the region of a given value of the independent variable (eg $\tan(\pi/4 + h) \approx 1 + 2h + 2h^2$, for small values of h)

Taylor's series
a power series that gives the expansion of a function f(x) in the neighbourhood of a point a provided that all derivatives exist and the series converges. If f(x) and its derivatives exist in the region of $x = a$, then

$$f(a + h) = f(a) + \frac{f'(a)}{1!}h + \frac{f''(a)}{2!}h^2 + \frac{f'''(a)}{3!}h^3 + \ldots$$

– compare MACLAURIN'S SERIES

Examples

(a) $\sin\left(\dfrac{\pi}{6} + h\right) = \dfrac{1}{2} + \dfrac{\sqrt{3}}{2}h - \dfrac{1}{4}h^2 + \dfrac{\sqrt{3}}{12}h^3 + \ldots$

(b) $\ln(1 + h) = h - \dfrac{1}{2}h^2 + \dfrac{1}{3}h^3 - \dfrac{1}{4}h^4 + \ldots$

t distribution
a probability density function that is used esp in testing whether a statistical sample is likely to have come from a larger sample of known statistical properties

term
an expression connected with another by a plus or minus sign (eg in the polynomial $x^4 - 7x^2 + 5x - 4$, there are four terms: x^4, $-7x^2$, $+5x$, and -4)

tessellation
the repeated use of a single shape to cover a plane surface

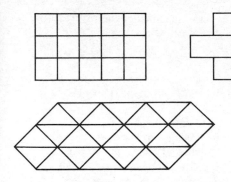

test
the procedure of submitting a statement to such conditions or operations as will lead to its proof or disproof or to its acceptance or rejection

tetra- *or* tetr-
four; having four; having four parts

tetrahedron
a polyhedron with four triangular faces

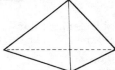

t-formulae
trigonometrical formulae which express sin θ, cos θ, and tan θ in terms of tan ½θ

$$\tan \theta = \frac{2t}{1 - t^2} \qquad \sin \theta = \frac{2t}{1 + t^2} \qquad \cos \theta = \frac{1 - t^2}{1 + t^2}$$

where $t = \tan \frac{1}{2}\theta$.

theorem
a statement, formula, or proposition in mathematics or logic deduced or to be deduced from other more basic formulae or propositions

theory
a body of theorems presenting a concise systematic view of a subject ⟨~ *of equations*⟩

theory of games
the analysis of a situation (eg in business or military strategy) in which opposing interests given specific information are allowed a

141

choice of moves with the object of maximizing their wins and minimizing their losses – called also GAME THEORY

theory of numbers
NUMBER THEORY

ton
an imperial unit of mass equal to 2240lb (approx 1016kg) – compare TONNE

tonne
a metric unit of mass equal to 1000kg – compare TON

topology
a branch of mathematics that deals with geometric properties which are unaltered by elastic deformation (eg stretching or twisting)

torque
the turning effect of a force or combination of forces – see MOMENT

torus
a ring-shaped surface or solid (eg a tyre inner tube) generated by a circle rotated about an axis in its plane that does not intersect the circle

total
1 a result of addition; a sum
2 an entire quantity; an amount

total differential
the sum of the products of each partial derivative of a function of several variables with the increment of the corresponding variable. If V is a function of x, y, and z then the total differential dV is given by

$$dV = \frac{\partial V}{\partial x}\,dx + \frac{\partial V}{\partial y}\,dy + \frac{\partial V}{\partial z}\,dz$$

trace
of a matrix the sum of the elements on the principal diagonal – called also SPUR

transcendental number
a number (eg e or π) that cannot be the root of a polynomial equation with rational coefficients – compare ALGEBRAIC NUMBER

transform (*noun*)
1 a mathematical element or term obtained from another by transformation
2 TRANSFORMATION

transform (*verb*)
to subject to mathematical transformation

transformation
1 the operation of changing (eg by rotation or mapping) one configuration or expression into another in accordance with a mathematical rule
2 the formula used for a transformation
– see REFLECTION, ROTATION, ENLARGEMENT, TRANSLATION, SHEAR

translate
to subject to mathematical translation

translation
transformation of a geometrical figure in which the figure moves without rotation. The size and shape of the figure are unchanged by a translation.

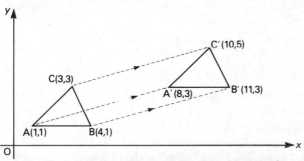

Triangle A′B′C′ is the image of triangle ABC after a translation $\binom{7}{2}$

All points in the figure move the same distance, along parallel lines – compare REFLECTION, ROTATION, ENLARGEMENT, SHEAR

transpose
a mathematical matrix formed by interchanging the rows of a given matrix with its corresponding columns
NOTATION The transpose of matrix **M** is written $\mathbf{M}^T$

    Example

$$\mathbf{M} = \begin{pmatrix} 1 & 2 & 3 \\ 4 & 5 & 6 \end{pmatrix} \qquad \mathbf{M}^T = \begin{pmatrix} 1 & 4 \\ 2 & 5 \\ 3 & 6 \end{pmatrix}$$

transversal
a line that crosses a system of lines

trapezium (*plural* **trapeziums** *or* **trapezia**)
a four-sided figure having two

trapezium rule

sides parallel. The area of a trapezium with parallel sides of lengths *a* and *b* separated by a perpendicular distance *h* is equal to ½*h*(*a* + *b*).

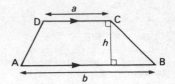

trapezium rule
a numerical method for estimating the area under a curve

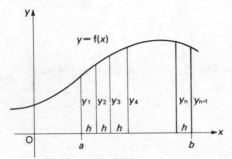

$$\int_a^b f(x)\,dx \approx \frac{1}{2}h\left(y_1 + 2y_2 + 2y_3 + 2y_4 + \ldots + 2y_n + y_{n+1}\right)$$

– compare SIMPSON'S RULE

tree diagram
in probability a treelike diagram in which the branches represent the possible outcomes of an experiment and their probabilities

Example A bag contains 3 red balls and 2 black balls. Two balls are removed at random and without replacement, calculate the probabilities of the possible outcomes.

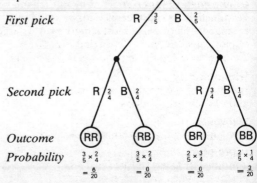

trend
the general movement in the course of time of a statistically detectable change; *also* a statistical curve reflecting such a change

trial
in statistics an instance of an experiment

triangle
a three-sided polygon – see also SINE RULE, COSINE RULE

triangle number
a member of the set of numbers 1, 3, 6, 10, 15, ... in which each number is obtained from its predecessor by adding the next natural number. Triangle numbers can be represented by triangular patterns of dots:

triangular matrix
a square matrix in which every element above (or below) the principal diagonal is zero, eg

$$\begin{pmatrix} 1 & 4 & 5 \\ 0 & 2 & 6 \\ 0 & 0 & 3 \end{pmatrix} \quad \begin{pmatrix} 1 & 0 & 0 \\ 4 & 2 & 0 \\ 5 & 6 & 3 \end{pmatrix}$$

triangulation
the measurement of the elements necessary to determine the network of triangles into which any part of the earth's surface is divided in surveying; *broadly* any trigonometric operation for finding a position or location by means of bearings from two fixed points a known distance apart

trigonometric function
1 a function (specif the sine, cosine, tangent, secant, cosecant, or cotangent) of an angle most simply expressed in terms of the ratios of pairs of sides of a right-angled triangle
2 the inverse (eg the arc sine) of a trigonometric function

trigonometry
the study of the properties of triangles and trigonometric functions and of their applications

trinomial
a polynomial of three terms (eg $2x^2 + 5x - 7$)

triple product
a product of three factors (eg $7 \times 11 \times 13$) – see SCALAR TRIPLE PRODUCT, VECTOR TRIPLE PRODUCT

trisect

to divide into three; *specif* to divide (an angle or line segment) into three equal parts

trivial

of or being the mathematically simplest case

trivial solution

a solution to an equation or set of equations in which all the variables are equal to zero – compare NONTRIVIAL

> **Example** The equations
> $$x + y + z = 0$$
> $$x + 2y + 3z = 0$$
> $$y + 2z = 0$$
> have a trivial solution $x = y = z = 0$,
> (but $x = 1$, $y = -2$, $z = 1$ is a *non*trivial one).

troy weight

the series of units of weight based on the pound of 12 ounces and the ounce of 20 pennyweights or 480 grains

true

1 logically necessary
2 corrected for error

truncate

to cut short:
1 *of a cone* to remove the top part (leaving a frustum)
2 *of an infinite series or decimal* to terminate the series or decimal after a finite number of terms (eg $\sin x \approx x - x^3/6$; $\pi \approx 3.14$)

truncation error

the error introduced into a calculation by truncating a series or a decimal

truth

the property (eg of a statement) of being in accord with fact or reality

truth table

a table that shows whether a compound statement is true or false in formal logic for each combination of truth-values of its component statements; *also* a similar table (eg for a computer logic circuit) showing the value of the output for each value of each input

> **Example**
>
> | A | ~A | B | ~B | A+B | AB |
> |---|----|----|----|-----|-----|
> | T | F | T | F | T | T |
> | T | F | F | T | T | F |
> | F | T | T | F | T | F |
> | F | T | F | T | F | F |

truth-value
the truth or falsity of a (logical) statement

t test
a statistical test using the t distribution

-tuple
set of (so many) elements – usu used with reference to mathematical sets with ordered elements ⟨*the ordered 2*-tuple *(a, b)*⟩

turning point
in calculus a point at which the gradient of a function is zero *and changes sign*. This may be a local maximum or a local minimum but *not* a point of inflection – compare STATIONARY POINT

In the diagram A and C are turning points, B is a point of inflection.

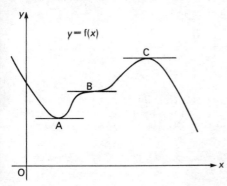

$y = f(x)$

two-tailed test *or* **two-tail test**
a statistical test of a hypothesis in which all the possible values of the test statistic (statistic used in testing a hypothesis) that would lead to acceptance of the hypothesis fall between two given values, with values greater than one of the given values or less than the other leading to rejection of the hypothesis – compare ONE-TAILED TEST

U

unbiased *or* **unbiassed**
of a statistic having an observed value equal to the expected value of the population parameter being estimated

union
of two sets A and B the set of all the elements which belong to A or B or both (written A ∪ B) – compare INTERSECTION

Example If A = {a, b, c, d} and B = {c, d, e}
then A ∪ B = {a, b, c, d, e}

unique
producing only one result (eg the equations $x + y = 4$ and $x - y = 2$, have a unique solution $x = 3$, $y = 1$)

unit
a standardized measure (eg of length, mass, or time) – see also SI UNITS, IMPERIAL UNIT

unit circle
a circle whose radius is one unit long

unit matrix
IDENTITY MATRIX

unit square
in transformation geometry the square whose vertices are at the points (0,0), (1,0), (1,1) and (0,1)

unit vector
a vector whose magnitude is one, esp **i, j,** and **k** – compare ZERO VECTOR

unity
1 the number one or a definite amount taken as one or that assumes the value of one for the purpose of calculation
2 a number by which any element of an arithmetical or mathematical system can be multiplied without change in the resultant value – see also ROOT OF UNITY

universal set (*symbol* **ε**)
the set containing *all* the elements relevant to a particular problem – compare EMPTY SET

unknown
a symbol in a mathematical equation representing an unknown quantity – see VARIABLE

upper bound

a number greater than or equal to every element of a given mathematical set (eg the upper bound of 0·3, 0·33, 0·333, 0·3333, … is ⅓)

V

valid
correctly derived from premises; logically sound ⟨~ *argument*⟩ ⟨~ *inference*⟩ ⟨~ *conclusion*⟩

value
1 a numerical quantity assigned or computed ⟨*the ~ of x is 95*⟩
2 the amount or extent of a specified measurement of space, time, or quantity

vanish
to assume the value zero

variable (*noun*)
(a symbol representing) a quantity that may assume any of a set of values. NB It is customary to use letters towards the end of the alphabet, esp from *t* onwards, to represent variables; letters at the beginning of the alphabet are normally used to represent constants. In either case, such letters are always printed in *italic* type.

variable (*adjective*)
having the characteristics of a variable ⟨*a ~ number*⟩

variance
the square of the standard deviation

variate
RANDOM VARIABLE

variation
1 the extent to which a thing varies
2 (a measure of) the change in the value of a mathematical variable or function
3 a relationship in which one quantity is proportional to another or the power of another (eg the area of a circle varies as the square of its radius).
y varies as x^n (written $y \propto x^n$) means that y and x are related by an equation of the form $y = kx^n$ (where k is a constant).
y varies *inversely* as x^n (written $y \propto 1/x^n$) means that y is related to x by an equation of the form $y = k/x^n$.

> **Example** The force with which the earth attracts an (external) object varies inversely as the square of the distance from the centre of the earth.

vector

The force F is related to the distance x, by an equation of the form

$$F = k/x^2$$

vector

1 vector, vector quantity a quantity (eg velocity or force) that has both magnitude and direction and that is usu represented by a straight line with an arrow whose length represents the magnitude and whose orientation in space represents the direction – compare SCALAR

2 an element of a vector space

vector product

a vector **c** whose length is the product of the lengths of two vectors **a** and **b** and the sine of their included angle, whose direction is that of a right-handed screw with axis **c** when **a** is rotated into **b**, and that is perpendicular to both **a** and **b** – compare SCALAR PRODUCT

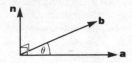

In the diagram, the vector product of **a** and **b** (written **a** ∧ **b**) is **ab** sin θ **n**, where **n** is a unit vector perpendicular to both **a** and **b**. The vector product is sometimes called the cross product and written **a** × **b**.

If $\mathbf{a} = x_1 \mathbf{i} + y_1 \mathbf{j} + z_1 \mathbf{k}$
and $\mathbf{b} = x_2 \mathbf{i} + y_2 \mathbf{j} + z_2 \mathbf{k}$
then

$$\mathbf{a} \wedge \mathbf{b} = \begin{vmatrix} \mathbf{i} & \mathbf{j} & \mathbf{k} \\ x_1 & y_1 & z_1 \\ x_2 & y_2 & z_2 \end{vmatrix}$$

vector quantity

VECTOR 1

vector space

a set whose elements are generalized vectors and which is a commutative group under addition that is also closed under an operation of multiplication by elements of a given field having the properties that

$$\lambda(\mathbf{a} + \mathbf{b}) = \lambda\mathbf{a} + \lambda\mathbf{b}$$
$$(\lambda + \mu)\mathbf{a} = \lambda\mathbf{a} + \mu\mathbf{a}$$
$$(\lambda\mu)\mathbf{a} = \lambda(\mu\mathbf{a})$$

where **a, b** are vectors and λ, μ are elements of the field

vector sum

the sum of vectors that for two vectors is geometrically represented by the diagonal of a parallelogram whose sides represent the two vectors being added

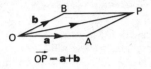

$$\overrightarrow{OP} = \mathbf{a} + \mathbf{b}$$

vector triple product

a vector quantity which is the product of three vectors **a**, **b** and **c** and is given by

$$\mathbf{a} \wedge (\mathbf{b} \wedge \mathbf{c}) = (\mathbf{a} . \mathbf{c})\mathbf{b} - (\mathbf{a} . \mathbf{b})\mathbf{c}$$

(NB It does *not* equal $(\mathbf{a} \wedge \mathbf{b}) \wedge \mathbf{c}$) – compare SCALAR TRIPLE PRODUCT

velocity

rate of change of displacement. The word *velocity* is used in preference to speed when *both* the magnitude *and the direction* of the motion are given or are to be calculated (eg a velocity of 500 km/h on a bearing of 270°), ie velocity is used for the vector quantity and speed for the scalar.

Venn diagram

a diagram that uses circles or other shapes to represent mathematical or logical relations between sets or propositions by the inclusion, exclusion, or intersection of the shapes [named after John Venn (1834–1923) Cambridge mathematician] – see also INTERSECTION, UNION, SYMMETRIC DIFFERENCE, UNIVERSAL SET

Example
$\varepsilon = \{1, 2, 3, 4, 5, 6, 7, 8, 9, 10\}$
$A = \{2, 4, 6, 8, 10\}$ and $B = \{3, 6, 9\}$

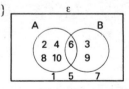

vertex (*plural* vertices)

1 the point opposite to and farthest from the base in a geometrical figure
2 the point from which an angle diverges or at which lines or curves intersect
3 a point where an axis of an ellipse, parabola, or hyperbola intersects the curve

vertical

1 perpendicular to the plane of the horizon or to an axis considered as horizontal
2 at a vertex ⟨vertical*ly opposite angles*⟩

vertically opposite angles
the (equal) angles formed when two lines intersect

volume
space occupied as measured in cubic units (eg litres); cubic
capacity ⟨*the ~ of a sphere*⟩

Examples

object	*dimensions*	*volume*
cube	side l	l^3
cuboid	length l, breadth b, height h	lbh
cone	base radius r, height h	$\frac{1}{3}\pi r^2 h$
pyramid	base area A, height h	$\frac{1}{3}Ah$
sphere	radius r	$\frac{4}{3}\pi r^3$

vulgar fraction
a fraction (eg ¾ of ½) in which both the denominator and
numerator are whole numbers – see also RATIONAL NUMBER,
IRRATIONAL NUMBER

Wallis' formulae *or* **Wallis's formulae**
a formula for finding the definite integral (between 0 and $\pi/2$) of an integral power of sin x

$$\int_0^{\pi/2} \sin^n x \ dx = \frac{(n-1)(n-3) \ \dots \ 3\times 1}{n \quad (n-2) \ \dots \ 4\times 2} \times \frac{\pi}{2} \quad \text{(if } n \text{ is even)}$$

$$= \frac{(n-1)(n-2) \ \dots \ 4\times 2}{n \quad (n-3) \ \dots \ 5\times 3} \quad \text{(if } n \text{ is odd)}$$

watt
the SI unit of power equal to the expenditure of 1 joule of energy in 1 second

weigh
1 to find the weight of an object
2 to have weight or a specified weight

weight (*noun*)
the force with which the earth attracts an object. If the mass is m kg and the gravitational acceleration is g ms^{-2}, then the weight is mg newtons.

weight (*verb*)
to assign a statistical weight to

weighted
1 having a statistical weight assigned
2 compiled from weighted statistical data

whole number
INTEGER

witch of Agnesi
a plane curve that is symmetrical about the y-axis, approaches the x-axis as an asymptote, and has the equation $x^2 y = 4a^2 (2a - y)$

work
work done by a force when the force and the displacement are in the same straight line = force x displacement. More generally, if the force is represented by vector $\mathbf{F}$ and the displacement by vector $\mathbf{x}$, the work done is the scalar product $\mathbf{F.x}$. In SI units work done is measured in joules.

x-axis
1 the horizontal axis in a Cartesian coordinate system (system of axes with reference to which points can be placed on a graph) having two axes at right angles
2 one of the three axes in a Cartesian coordinate system having three axes at right angles

x-coordinate
a coordinate that defines the position of a point (eg on a graph) and whose value is determined by measuring along the direction of an *x*-axis; *esp* ABSCISSA

x-intercept
the *x*-coordinate of a point (eg on a graph) where a line, curve, or surface intersects the *x*-axis

yard
an imperial unit of length equal to 3ft (about 0·9144m)

y-axis
1 the axis perpendicular to the *x*-axis in a Cartesian coordinate system (system of axes with reference to which points can be placed on a graph) having two axes at right angles
2 one of the three axes in a Cartesian coordinate system having three axes at right angles

y-coordinate
a coordinate that defines the position of a point (eg on a graph) and whose value is determined by measuring along the direction of a *y*-axis; *esp* ORDINATE

y-intercept
the *y*-coordinate of a point (eg on a graph) where a line, curve, or surface intersects the *y*-axis

Z

z-axis
the axis in a three-dimensional
system that is perpendicular to
both the *x*- and *y*-axes.
(Normally the *x*- and *y*-axes are
taken to be in the horizontal
plane, and the *z*-axis vertical;
the *x*-, *y*- and *z*-axes should
form a right-handed set.)

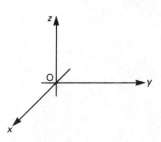

zero (*plural* **zeros** *also* **zeroes**)
1 the arithmetical symbol 0 or $\emptyset$ denoting the absence of all
magnitude or quantity
2 an identity element of a group under addition
3 a value of the independent variable of a function that makes it
equal to zero

zero matrix
a square matrix with all its entries zero, eg $\begin{pmatrix} 0 & 0 & 0 \\ 0 & 0 & 0 \\ 0 & 0 & 0 \end{pmatrix}$

– compare IDENTITY MATRIX

zero vector
a vector of zero magnitude – compare UNIT VECTOR

zone
a portion of the surface of a sphere included between two parallel
planes. Area of the zone = $2\pi r x$ where r is the radius of the
sphere and x is the distance between the planes.

Mathematical notation

1 Set notation

$\in$	is an element of.
$\notin$	is not an element of.
$\{a, b, c, ...\}$	the set with elements $a, b, c, ...$
$\{x: ...\}$	the set of elements x, such that ...
$n(A)$	the number of elements in set A.
$\varnothing$	the empty set.
ε	the universal set.
A'	the complement of set A.
$\mathbb{N}$	the set of natural numbers (including zero) $0, 1, 2, 3 ...$
$\mathbb{Z}$	the set of integers $0, \pm 1, \pm 2, \pm 3 ...$
$\mathbb{Z}^+$	the set of positive integers $+1, +2, +3 ...$
$\mathbb{Q}$	the set of rational numbers.
$\mathbb{R}$	the set of real numbers.
$\mathbb{C}$	the set of complex numbers.
$\subseteq$	is a subset of.
$\subset$	is a proper subset of.
$\cup$	union.
$\cap$	intersection.
$[a, b]$	the closed interval $\{x \in \mathbb{R}: a \leq x \leq b\}$.
(a, b)	the open interval $\{x \in \mathbb{R}: a < x < b\}$.

2 Miscellaneous symbols

$=$	is equal to.
$\neq$	is not equal to.
$>, <$	is greater than, is less than.
$\geq, \leq$	is greater than or equal to, is less than or equal to.
$\approx$	is approximately equal to.

3 Operations

$a + b$	a plus b.
$a - b$	a minus b.
$a \times b, ab, a.b$	a multiplied by b.
$a \div b, \dfrac{a}{b}, a/b$	a divided by b.
$\displaystyle\sum_{i=1}^{i=n} a_i$	$a_1 + a_2 + a_3 + ... + a_n$.

4 Functions

f(x)	the value of the function f at x.
f: $A \rightarrow B$	f is a function which maps each element of set A onto a member of set B.
f: $x \mapsto y$	f maps the element x onto an element y.
f^{-1}	the inverse of the function f.
g∘f or gf	the composite function g(f(x)).
$\lim_{x \to a}$ f(x)	the limit of f(x) as x tends to a.
δx	an increment of x.
$\dfrac{dy}{dx}$	the derivative of y with respect to x.
$\dfrac{d^n y}{dx^n}$	the nth derivative of y with respect to x.
f'(x), f''(x), ... f$^{(n)}$(x)	the first, second, ... nth derivatives of f(x).
$\int y \, dx$	the indefinite integral of y with respect to x.
$\int_a^b y \, dx$	the definite integral, with limits a and b.
$\left[F(x) \right]_a^b$	$F(b) - F(a)$.

5 Exponential and logarithmic functions

e^x or exp$x$	the exponential function.
$\log_a x$	logarithm of x in base a logarithms.
$\ln x$	$\log_e x$.
$\lg x$	$\log_{10} x$.

6 Circular and hyperbolic functions

$\sin x$, $\cos x$, $\tan x$	the circular functions sine, cosine, tangent.
$\operatorname{cosec} x$, $\sec x$, $\cot x$	the reciprocals of the above functions.
$\sin^{-1} x$ or $\arcsin x$	the inverse of the function $\sin x$ (with similar abbreviations for the inverses of the other circular functions).
$\sinh x$ etc.	the hyperbolic functions.

7 Other functions

$\sqrt{a}$	the positive square root of a.
$|a|$	the modulus of a.
$n!$	n factorial; $n! = n \times (n-1) \times (n-2) \times \ldots \times 3 \times 2 \times 1.$ $(0! = 1).$
$\dbinom{n}{r}$	$\dfrac{n!}{r!(n-r)!}$ when $n, r \in \mathbb{N}$ and $0 \leqslant r \leqslant n$.

$$\binom{n}{r} \qquad \frac{n(n-1) \dots (n-r+1)}{r!}$$

when $n \in \mathbb{Q}$ and $r \in \mathbb{N}$.

8 Complex numbers

i	the square root of -1.
z or w	a typical complex number, e.g. $x + \mathrm{i}y$, where $x, y \in \mathbb{R}$.
$\mathrm{Re}(z)$	the real part of z; $\mathrm{Re}(x + \mathrm{i}y) = x$.
$\mathrm{Im}(z)$	the imaginary part of z; $\mathrm{Im}(x + \mathrm{i}y) = y$.
$\lvert z \rvert$	the modulus of z; $\lvert x + \mathrm{i}y \rvert = \surd(x^2 + y^2)$.
$\arg(z)$	the argument of z.
z^*	the complex conjugate of z.

9 Matrices

$\mathbf{M}$	a typical matrix $\mathbf{M}$.
$\mathbf{M}^{-1}$	the inverse of a matrix $\mathbf{M}$ (provided it exists).
$\mathbf{M}^{\mathrm{T}}$	the transpose of matrix $\mathbf{M}$.
$\det(\mathbf{M})$	the determinant of a square matrix $\mathbf{M}$.
$\mathrm{adj}(\mathbf{M})$	the adjoint of a square matrix $\mathbf{M}$.
$\mathbf{I}$	the identity matrix.

10 Vectors

$\mathbf{a}$	the vector $\mathbf{a}$.
$\lvert \mathbf{a} \rvert$ or a	the magnitude of vector $\mathbf{a}$.
$\hat{\mathbf{a}}$	the unit vector with the same direction as $\mathbf{a}$.
$\mathbf{i}, \mathbf{j}, \mathbf{k}$	unit vectors parallel to the Cartesian coordinate axes.
$\overrightarrow{\mathrm{AB}}$	the vector represented by the line segment AB.
$\lvert \overrightarrow{\mathrm{AB}} \rvert$ or AB	the length of the vector $\overrightarrow{\mathrm{AB}}$.
$\mathbf{a}.\mathbf{b}$	the scalar product of $\mathbf{a}$ and $\mathbf{b}$.
$\mathbf{a} \wedge \mathbf{b}$	the vector product of $\mathbf{a}$ and $\mathbf{b}$.

Metric prefixes

Prefix	Abbreviation	Factor
mega	M	10^6
kilo	k	10^3
hecto	h	10^2
deca	da	10^1
deci	d	10^{-1}
centi	c	10^{-2}
milli	m	10^{-3}
micro	μ	10^{-6}
nano	n	10^{-9}
pico	p	10^{-12}

Approximate conversions (to 3 significant figures)

Imperial	Metric	Metric	Imperial
1 inch	2·54 centimetre	1 cm	0·394 in
1 foot	0·305 metre	1 m	3·28 ft
1 yard	0·914 metre	1 m	1·09 yd
1 mile	1·61 kilometre	1 km	0·621 mile
1 acre	0·405 hectare	1 ha	2·47 acres
1 pound	0·454 kilogram	1 kg	2·20 lb
1 ton	1·016 tonne	1 t	0·984 tons
1 pint	0·568 litre	1 l	1·76 pints
1 gallon	4·55 litres	1 l	0·220 gal

Useful constants (correct to 6 significant figures)

$\pi = 3·14159$ $\qquad\qquad$ $e = 2·71828$

Physical constants (correct to 3 significant figures)

$g = 9·81$ m/s^2

velocity of sound $= 332$ m/s (in dry air at 0°C)

velocity of light $= 3·00 \times 10^5$ km/s